Foundations of Systems Biology

Foundations of Systems Biology

edited by
Hiroaki Kitano

The MIT Press
Cambridge, Massachusetts
London, England

This book was set in Palatino by the author using the LaTeX document preparation system. Printed on recycled paper and bound in the United States of America.

Library of Congress Cataloging-in-Publication Data

Foundations of systems biology
edited by Hiroaki Kitano. - 1st ed.
p.cm.
Included bibliographical references (p.).
ISBN 0-262-11266-3 (hc.: alk. paper)
1. Biological systems-Research-Methodology
2. Biological systems-Research-Case studies.
I. Kitano, Hiroaki, 1961-

QH313.F662002
573-dc21

2001042807

Contents

Contributors

Mutsuki Amano
Division of Signal Transduction,
Nara Institute of Science and Technology.

Katja Bettenbrock
bettenbrock@mpi-magdeburg.mpg.de
Max-Planck-Institute for Dynamics of Complex Technical Systems.

Hamid Bolouri
hbolouri@caltech.edu
JST/ERATO Kitano Systems Biology Project,
and
Control and Dynamical Systems,
California Institute of Technology,
and
Division of Biology,
California Institute of Technology,
and
Science and Technology Research Centre,
University of Hertfordshire.

Dennis Bray
d.bray@zoo.cam.ac.uk
Department of Zoology,
University of Cambridge.

Jehoshua Bruck
bruck@paradise.caltech.edu
Division of Engineering and Applied Science,
California Institute of Technology.

John Doyle
doyle@cds.caltech.edu
Kitano Systems Biology Project,
ERATO, JST,
and
Control and Dynamical Systems,
California Institute of Technology.

Andrew Finney
afinney@cds.caltech.edu
JST/ERATO Kitano Systems Biology Project,
and
Control and Dynamical Systems,
California Institute of Technology.

Ernst Dieter Gilles
gilles@mpi-magdeburg.mpg.de
Max-Planck-Institute for Dynamics of Complex Technical Systems.

Martin Ginkel
ginkel@mpi-magdeburg.mpg.de
Max-Planck-Institute for Dynamics of Complex Technical Systems.

Shugo Hamahashi
shugo@symbio.jst.go.jp
Kitano Symbiotic Systems Project,
ERATO, JST,
and
Department of Computer Science,
Keio University.

Michael Hucka
mhucka@cds.caltech.edu
JST/ERATO Kitano Systems Biology Project,
and
Control and Dynamical Systems,

California Institute of Technology.

Kozo Kaibuchi
Division of Signal Transduction,
Nara Institute of Science and Technology,
and
Department of Cell Pharmacology,
Nagoya University.

Mitsuo Kawato
Kawato Dynamic Brain Project,
ERATO, JST,
and
Human Information Processing Research Laboratories,
ATR.

Martin A. Keane
makeane@ix.netcom.com
Econometric Inc.

Hiroaki Kitano
kitano@symbio.jst.go.jp
Kitano Symbiotic Systems Project,
ERATO, JST,
and
The Systems Biology Institute,
and
Control and Dynamical Systems,
California Institute of Technology,
and
Sony Computer Science Laboratories, Inc.

John R. Koza
koza@stanford.edu
Biomedical Informatics,
Department of Medicine,
Department of Electrical Engineering,
Stanford University.

Andreas Kremling
kre@mpi-magdeburg.mpg.de
Max-Planck-Institute for Dynamics of Complex Technical Systems.

Shinya Kuroda
kuroda@fido.cpmc.columbia.edu
Kawato Dynamic Brain Project,
ERATO, JST,
and
Division of Signal Transduction,
Nara Institute of Science and Technology,
Present address: Center for Neurobiology and Behavior,
Columbia University.

Koji M. Kyoda
kyoda@symbio.jst.go.jp
Kitano Symbiotic Systems Project,
ERATO, JST,
and
Department of Fundamental Science and Technology,
Keio University.

Guido Lanza
guido@pharmix.com
Genetic Programming Inc.

Andre Levchenko
andre@paradise.caltech.edu
Division of Engineering and Applied Science,
California Institute of Technology.

Pedro Mendes
mendes@vt.edu
Virginia Bioinformatics Institute,
Virginia Polytechnic Institute and State University.

Satoru Miyano
miyano@ims.u-tokyo.ac.jp
Kitano Symbiotic Systems Project,
ERATO, JST,
and
Human Genome Center, Institute of Medical Science,
University of Tokyo.

Control and Dynamical Systems,
California Institute of Technology.

Eric Mjolsness
mjolsness@jpl.nasa.gov
Jet Propulsion Laboratory,
California Institute of Technology,
and
Division of Biology,
California Institute of Technology,
and
Kitano Symbiotic Systems Project,
ERATO, JST.

Mineo Morohashi
moro@symbio.jst.go.jp
Kitano Symbiotic Systems Project,
ERATO, JST,
and
Department of Fundamental
Science and Technology,
Keio University.

William Mydlowec
myd@cs.stanford.edu
Genetic Programming Inc.

Masao Nagasaki
masao@ims.u-tokyo.ac.jp
Kitano Symbiotic Systems Project,
ERATO, JST,
and
Department of Information
Science,
Human Genome Center, Institute
of Medical Science,
University of Tokyo.

Yoichi Nakayama
ynakayam@sfc.keio.ac.jp
Institute for Advanced Biosciences,
Keio University.

Shuichi Onami
sonami@symbio.jst.go.jp
Kitano Symbiotic Systems Project,
ERATO, JST,
and
The Systems Biology Institute,
and

Herbert Sauro
hsauro@cds.caltech.edu
JST/ERATO Kitano Systems
Biology Project,
and
Control and Dynamical Systems,
California Institute of Technology.

Nicolas Schweighofer
Kawato Dynamic Brain Project,
ERATO, JST,
Present address: Learning Curve.

Bruce Shapiro
bshapiro@jpl.nasa.gov
Jet Propulsion Laboratory,
California Institute of Technology.

Thomas Simon Shimizu
tss26@cam.ac.uk
Department of Zoology,
University of Cambridge.

Jörg Stelling
stelling@mpi-magdeburg.mpg.de
Max-Planck-Institute for
Dynamics of Complex Technical
Systems.

Paul W. Sternberg
pws@its.caltech.edu
Division of Biology and
Howard Hughes Medical Institute,
California Institute of Technology.

Zoltan Szallasi
zszallas@mxc.usuhs.mil
Department of Pharmacology,
Uniformed Services University of
the Health Sciences.

Masaru Tomita
mt@sfc.keio.ac.jp
Institute for Advanced Biosciences,

Keio University.

Mattias Wahde
mwahde@me.chalmers.se
Division of Mechatronics,
Chalmers University of
Technology.

Tau-Mu Yi
tmy@caltech.edu
Kitano Symbiotic Systems Project,
ERATO, JST,
and
Division of Biology,
California Institute of Technology.

Jessen Yu
jyu@cs.stanford.edu
Genetic Programming Inc.

Preface

Systems biology aims at understanding biological systems at system level. It is a growing area in biology, due to progress in several fields. The most critical factor has been rapid progress in molecular biology, furthered by technologies for making comprehensive measurements on DNA sequence, gene expression profiles, protein-protein interactions, etc. With the ever-increasing flow of biological data, serious attempts to understand biological systems as systems are now almost feasible. Handling this high-throughput experimental data places major demands on computer science, including database processing, modeling, simulation, and analysis. Dramatic progress in semiconductor technologies has led to high-performance computing facilities that can support system-level analysis.

This is not the first attempt at system-level analysis; there have been several efforts in the past, the most notable of which is cybernetics, or biological cybernetics, proposed by Norbert Wiener more than 30 years ago. With the limited understanding of biological processes at the molecular level at that time, most of the work was on phenomenological analysis of physiological processes. There have also been biochemical approaches, such as metabolic control analysis, and although restricted to steady-state flow, it has successfully been used to explore system-level properties of biological metabolism. Systems biology, just like all other emerging scientific disciplines, is built on multiple efforts that share the vision. However, systems biology is distinct from past attempts because for the first time we are able to understand biology at the system level based on molecular-level understanding, and to create a consistent system of knowledge grounded in the molecular level. In addition, it should be noted that systems biology is intended to be biology for system-level studies, not physics, systems science, or informatics, which try to apply certain dogmatic principles to biology.

When the field has matured in the next few years, systems biology will be characterized as a field of biology at the system level with extensive use of cutting-edge technologies and highly automated high-throughput precision measurement combined with sophisticated computational tools and analysis. Systems biology clearly includes both experimental and computational or analytical studies. However, systems biology is not a mere combination of molecular biology and computing science to reverse engineer gene networks. Of course, it is one of the topics included, but system-level understanding requires more than understanding structures. Understanding of (1) system structures, (2) dynamics, (3) control methods, and (4) design methods are major four milestones of systems biology research. One of the most important missions of this book, which I tried hard in my chapter, is to define the scope and provide the vision and perspectives of this new born field.

I am very pleased to see that interest is rapidly growing among both experimental biologists and those who with computing and engineering backgrounds are seriously interested in biological systems. Not many people understood what I was trying to describe when I was using the term "systems biology" a few years ago, because it was well before human genome sequence to complete, high-throughput experiments were to be considered as realistic option, and it was a new term nobody used before. But, today more and more people are using the concept and the term. Of course, it is the actual research that matters, but the term is also important because it symbolically represents what we are trying to accomplish. Today, we find more and more researchers are getting involved, as well as numbers of research groups and institutions are being formed focusing on systems biology.

Fortunately, I managed to convince the Japanese government to support the initiation of a new international conference. The First International Conference on Systems Biology (ICSB2000) was held in Tokyo from November 14–16, 2000, supported by the Japan Science and Technology Corporation, an agency belonging to the Science and Technology Agency of the Japanese government. It was the first international conference that clearly focused on systems biology work. Since then, various international and national conferences, symposiums, and seminars have started organize systems biology sessions. The second conference will be held at the California Institute of Technology in 2001 with the support of Caltech. In fact, Caltech is one of the first institutions that seriously explored systems biology. I still remember the overwhelming reaction when I gave a talk "Perspectives on Systems Biology" at Caltech in March 1998.

This book is the first book on systems biology, and consists of papers representing work in the systems biology field. It is loosely based on papers that were presented at ICSB2000. Of course, many research studies related to systems biology are already underway, and I must state that this book is by no means an exhaustive collection of such works. Also, the experimental aspects of systems biology are under-represented here, because many of the projects aiming at next-generation experiments are at their early stage and so are not ready for publication. Nevertheless, the book covers the central themes of systems biology: comprehensive and automated measurements, reverse engineering of gene and metabolic networks from experimental data, software issues, modeling and simulation, and system-level analysis.

Although it it based on a long history, systems biology is a field in its infancy. This book serves two purposes: first, to inform interested researchers on the current state of the research and challenges before us, and second, to be an archival collection of papers to record the initial stage of the research. It is likely, just as in any fast-growing research area, that the technical contents of the book will quickly become obsolete. However, it is often the case that the vision, concept, and philosophy are still valid and add value. I hope the field will quickly grow and flourish beyond its present boundaries, but that the vision outlined herein is enduring.

Finally, this book could not have been completed without the support of many people. Mineo Morohashi has done a beautiful job in sorting out and formatting all papers, communicating with authors as well as with The MIT Press, and many other tasks. While I was preoccupied with establishing the new research institution, The Systems Biology Institute (http://www.systems-biology.org/), he did most of the editorial assistant work for me. Thank you, Mineo. Members of the systems biology group of ERATO Kitano Symbiotic Systems Project have been a great help in soliciting papers and, more important, in formulating the basic concepts and vision behind systems biology. John Doyle, Mel Simon, Hamid Bolouri, and Mark Borisuk have been particularly cooperative and supportive. Mario Tokoro and Toshi Doi provided me with a superb research environment at Sony Computer Science Laboratories, Inc. Bob Prior at The MIT Press supported me in this project from the beginning; it was in the summer of 1997 at the International Joint Conference on Artificial Intelligence (IJCAI-97) in Nagoya that Bob walked up to me with a printout of the web page from a talk I had given on systems biology at the University of Cambridge, and asked me to publish this book. I am deeply indebted to all of you.

Hiroaki Kitano
Senior Researcher, Sony Computer Science Laboratories, Inc.
Director, ERATO Kitano Symbiotic Systems Project, JST, and
Director, The Systems Biology Institute
Tokyo, Japan

1 Systems Biology: Toward System-level Understanding of Biological Systems

Hiroaki Kitano

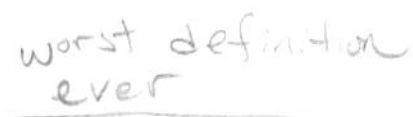

Systems biology is a new field in biology that aims at system-level understanding of biological systems. While molecular biology has led to remarkable progress in our understanding of biological systems, the current focus is mainly on identification of genes and functions of their products which are components of the system. The next major challenge is to understand at the system level biological systems that are composed of components revealed by molecular biology. This is not the first attempt at system-level understanding, since it is a recurrent theme in the scientific community. Nevertheless, it is the first time that we may be able to understand biological systems grounded in the molecular level as a consistent framework of knowledge. Now is a golden opportunity to uncover the essential principles of biological systems and applications backed up by in-depth understanding of system behaviors. In order to grasp this opportunity, it is essential to establish methodologies and techniques to enable us to understand biological systems in their entirety by investigating: (1) the structure of the systems, such as genes, metabolism, and signal transduction networks and physical structures, (2) the dynamics of such systems, (3) methods to control systems, and (4) methods to design and modify systems for desired properties. This chapter gives an overview of the field of systems biology that will provide a system-level understanding of life.

INTRODUCTION

The ultimate goal of biology is to understand every detail and principle of biological systems. Almost fifty years ago, Watson and Crick identified the structure of DNA (Watson and Crick, 1953), thus revolutionizing the way biology is pursued. The beauty of their work was that they grounded biological phenomena on a molecular basis. This made it possible to describe every aspect of biology, such as heredity, development, disease, and evolution, on a solid theoretical ground. Biology became part of a consistent framework of knowledge based on fundamental laws of physics.

Since then, the field of molecular biology has emerged and enormous progress has been made. Molecular biology enables us to understand biological systems as molecular machines. Today, we have in-depth understanding of elementary processes behind heredity, evolution, development, and disease. Such mechanisms include replication, transcription, translation, and so forth.

Large numbers of genes and the functions of their transcriptional products have been identified, with the symbolic accomplishment of the complete sequencing of DNA. DNA sequences have been fully identified for various organisms such as *mycoplasma, Escherichia coli (E. coli), Caenorhabditis elegans (C. elegans), Drosophila melanogaster,* and *Homo sapiens*. Methods to obtain extensive gene expression profiles are now available that provide comprehensive measurement at the mRNA level. Measurement of protein level and their interactions is also making progress (Ito et al., 2000; Schwikowski et al., 2000). In parallel with such efforts, various methods have been invented to disrupt the transcription of genes, such as loss-of-function knockout of specific genes and RNA interference (RNAi) that is particularly effective for *C. elegans* and is now being applied for other species.

There is no doubt that our understanding of the molecular-level mechanisms of biological systems will accelerate. Nevertheless, such knowledge does not provide us with an understanding of biological systems as systems. Genes and proteins are components of the system. While an understanding of what constitutes the system is necessary for understanding the system, it is not sufficient.

Systems biology is a new field of biology that aims to develop a system-level understanding of biological systems (Kitano, 2000). System-level understanding requires a set of principles and methodologies that links the behaviors of molecules to system characteristics and functions. Ultimately, cells, organisms, and human beings will be described and understood at the system level grounded on a consistent framework of knowledge that is underpinned by the basic principles of physics.

It is not the first time that system-level understanding of biological system has been pursued; it is a recurrent theme in the scientific community. Norbert Wiener was one of the early proponents of system-level understanding that led to the birth of cybernetics, or biological cybernetics (Wiener, 1948). Ludwig von Bertalanffy proposed general system theory (von Bertalanffy, 1968) in 1968 in an attempt to establish a general theory of the system, but the theory was too abstract to be well grounded. A precursor to such work can be found in the work of Cannon, who proposed the concept of "homeostasis" (Cannon, 1933). With the limited availability of knowledge from molecular biology, most such attempts have focused on the description and analysis of biological systems at the physiological level. The unique feature of systems biology that distinguishes it from past attempts is that there are opportunities to ground system-level understanding directly on the molecular level such as genes and proteins, whereas past attempts have not been able to sufficiently connect system-level description to molecular-level knowledge. Thus, although it is not the first time that system-level understanding has been pursued, it is the first time to have an opportunity to understand biological systems within the consistent framework of knowledge built up from the molecular level to the system level.

The scope of systems biology is potentially very broad and different sets of techniques may be deployed for each research target. It requires collective efforts from multiple research areas, such as molecular biology, high-precision measurement, computer science, control theory, and other scientific and engineering fields. Research needs to be carried out in four key areas: (1) genomics and other molecular biology research, (2) computational studies, such as simulation, bioinformatics, and software tools, (3) analysis of dynamics of the system, and (4) technologies for high-precision, comprehensive measurements.

This constitutes a major multi-disciplinary research effort that will enable us to understand biological systems as systems. But what does this mean? "System" is an abstract concept in itself. It is basically an assembly of components in a particular formation, yet it is more than a mere collection of components. To understand the system, it is essential that it can be not only to describe in detail, but also it to comprehend what happens when certain stimuli or disruptions occur. Ultimately, we should be able to design the system to meet specific functional properties. It takes more than a simple in-depth description; it requires more active synthesis to ensure that we have fully understood it.

To be more specific, in order to understand biological systems as systems, we must accomplish the following.

System Structure Identification: First of all, the structures of the system need to be identified, primarily such as regulatory relationships of genes and interactions of proteins that provide signal transduction and metabolism pathways, as well as the physical structures of organisms, cells, organella, chromatin, and other components.
Both the topological relationship of the network of components as well as parameters for each relation need to be identified. The use of high-throughput DNA microarray, protein chips, RT-PCR, and other methods to monitor biological processes in bulk is critical. Nevertheless, methods to identify genes and metabolism networks from these data have yet to be established.
Identification of gene regulatory networks[1] for multicellular organisms is even more complex as it involves extensive cell-cell communication and physical configuration in three-dimensional space. Structure identification for multicellular organisms inevitably involves not only identifying the structure of gene regulatory networks and metabolism networks, but also understanding the physical structures of whole animals precisely at the

1 In this article, the term "gene regulatory networks" is used to represent networks of gene regulations, metabolic pathways, and signal transduction cascades.

cellular level. Obviously, new instrumentation systems need to be developed to collect necessary data.

System Behavior Analysis: Once a system structure is identified to a certain degree, its behavior needs to be understood. Various analysis methods can be used. For example, one may wish to know the sensitivity of certain behaviors against external perturbations, and how quickly the system returns to its normal state after the stimuli. Such an analysis not only reveals system-level characteristics, but also provides important insights for medical treatments by discovering cell response to certain chemicals so that the effects can be maximized while lowering possible side effects.

System Control: In order to apply the insights obtained by system structure and behavior understanding, research into establishing a method to control the state of biological systems is needed. How can we transform cells that are malfunctioning into healthy cells? How can we control cancer cells to turn them into normal cells or cause apoptosis? Can we control the differentiation status of a specific cell into a stem cell, and control it to differentiate into the desired cell type? Technologies to accomplish such control would enormously benefit human health.

System Design: Ultimately, we would like to establish technologies that allow us to design biological systems with the aim of providing cures for diseases. One futuristic example would be to actually design and grow organs from the patient's own tissue. Such an organ cloning technique would be enormously useful for the treatment of diseases that require organ transplants. There may be some engineering applications by using biological materials for robotics or computation. By using materials that have self-repair and self-sustaining capability, industrial systems will be revolutionized.

This chapter discusses scientific and engineering issues to accomplish in-depth understanding of the system.

MEASUREMENT TECHNOLOGIES AND EXPERIMENTAL METHODS

Toward Comprehensive Measurements

A comprehensive data set needs to be produced to grasp an entire picture of the organism of interest. For example, the entire sequence has been deduced for yeast, and a microarray that can measure the expression level of all known genes is readily available. In addition, extensive studies of protein-protein interactions using the two-hybrid method are being carried out (Ito et al., 2000; Schwikowski et al., 2000). Efforts to obtain high-resolution spatiotemporal localization data for protein are underway.

C. elegans is an example of an intensively measured multi-cellular organism. A complete cell lineage has already been identified (Sulston et al., 1983; Sulston and Horvitz, 1977), the topology of the neural system has been fully described (White et al., 1986), the DNA sequence has been fully identified (The C. elegans Sequencing Consortium, 1998), a project for full description of gene expression patterns during development using whole-mount *in situ* hybridization (Tabara et al., 1996) is underway, and the construction of a systematic and exhaustive library of mutants has begun. In addition, a series of new projects has started for measuring neural activity *in vivo*, and for automatic construction of cell lineage in real time using advanced image processing combined with special microscopy (Yasuda et al., 1999; Onami et al., 2001a).

While yeast and *C. elegans* are examples of comprehensive and exhaustive understanding of biological systems, similar efforts are now being planned for a range of biological systems. Although these studies are currently limited to understanding the components of the system and their local relationship with other components, the combination of such exhaustive experimental work and computational and theoretical research would provide a viable foundation for systems biology.

Measurement for Systems Biology

Although efforts to systematically obtain comprehensive and accurate data sets are underway, systems biology is much more demanding for experimental biologists than the current practice of biology. It requires a comprehensive body of data and control of the quality of data produced so that it can be used as a reference point of simulation, modeling, and system identification. Eventually, many of the current experimental procedures must be automated to enable high-throughput experiments to be carried out with precise control of quality. Needless to say, not all biological experiments will be carried out in such an automated fashion, for important contributions will be made by small-scale experiments. Nevertheless, large-scale experiments will lay the foundation for system-level understanding.

High-throughput, comprehensive, and accurate measurement is the most essential part of biological science. While expectations are high for a computational approach to overcome limitations in the traditional approach in biology, it will never generate serious results without experimental data upon which computational studies can be grounded. For the computational and systems approach to be successful, measurement has to be (1) comprehensive, (2) quantitatively accurate, and (3) systematic.

While the requirement for quantitative accuracy is obvious, the other two criteria need further clarification. Comprehensiveness can be further classified into three types:

Factor comprehensiveness: Comprehensiveness in terms of target factors that are being measured, such as numbers of genes and proteins. It is important that measurement is carried out intensively for the factors (genes and proteins) that are related to the central genes and proteins of interest. Unless all genes and proteins are measured, how effectively measurement covers the factors of interest is more important, rather than the sheer number of factors measured.

Time-series comprehensiveness: In modeling and analysis of a dynamical system, it is important to capture its behavior with fine-grain time series. Traditional biological experiments tend to measure only the change before and after a certain event. For computational analysis, data measured at a constant time interval are essential in addition to traditional sampling points.

Item comprehensiveness: There are cases where several features, such as transcription level, protein interaction, phosphorylation, localization, and other features, have to be measured intensively for the specific target.

"Systematic" means that measurement is performed in such a way that obtained data can be consistently integrated. The ideal systematic measurement is simultaneous measurement of multiple features for a single sample. It is not sufficient to develop a sophisticated model and perform analysis using only the mRNA or protein level. Multiple data need to be integrated. Then, each data point has to be obtained using samples that are consistent across various measurements. If samples are prepared in substantially different ways, two data points cannot be integrated. Although this requirement sounds obvious, very few data sets meet these criteria today.

These criteria are elucidated in the scenario below with some examples of requirements for experimental data.

For example, to infer genetic regulatory networks from an expression profile, comprehensive measurement of the gene expression profile needs to be carried out. Expression data in which only the wild-type is measured is generally unusable for this purpose. The data should have a comprehensive set of deletion mutant and overexpression of each gene. Desirable data sets knock out all genes that are measured in the microarray. If only a limited number of genes can be knocked out due to cost and time constraints, it is critical that genes that are expected to be tightly coupled are intensively knocked out rather than knocking out genes sparsely over the whole possible regulatory network. This is due to computational characteristics of the reverse engineering algorithm that constructs the gene regulatory network from profile data. With such algorithms, sparse data points leave almost unlimited ambiguities on possible network structures. Even with the same number of data points, the algorithm produces much more reliable network hypotheses if measured genes are closely related. This is what is meant by factor comprehensiveness.

Time-series comprehensiveness is required for phenomena that are time aligned. Time-series profile data need to be prepared with particular caution in terms of time synchronization of samples to be measured.

It is often the case in traditional experiments that only two measurement points are set: one before the event and one after the event. For example, many studies in cellular aging research measured the expression level of aging-related genes for young cells, aged cells, and immortalized cells, without measuring changes of expression level on fine-grain time series. In some cases, time-series changes of expression level can be important information to create candidate hypotheses or eliminate possible mechanisms. In addition to measurements before and after a biologically interesting event, measurement should be carried at a constant time interval. Expression profile data that has reliable sample time synchrony and constant time interval is most useful to enable the computational algorithm to reliably fit models and parameters to experimental data.

Additional information from protein-protein interactions, such as from yeast two-hybrid experiments, is very useful to infer protein-level interactions that fill the gap between regulation of genes. Both protein interactions and expression profiles should be measured on samples that are prepared identically. This systematic measurement requirement is rather hard to meet currently, because not many research groups are proficient in multiple measurement techniques.

After obtaining gene regulatory networks, one needs to find out specific parameters used in the network. To understand dynamics, it is essential that each parameter regarding the network is obtained, so that various numerical simulations and analyses can be performed. Such parameters are binding constant, transcription rate, translation rate, chemical reaction rate, degradation rate, diffusion rate, speed of active transport, etc. Except for special cases, such as red blood cells, these constants are not readily available. Measurement using extracts provides certain information, but often these rate constants vary drastically *in vivo*. Ideally, comprehensive measurement of major parameters would be performed *in vivo*, but any measurement that gives reasonable estimates would be of great help. In addition to parameter measurement, it is critically important to measure the phosphorylation state at high resolution.

While accuracy is important, the level of accuracy required may vary depending on which part of the system is to be measured. In some parts of the network, the system behavior is sensitive to specific parameter values, and thus has to be measured with high accuracy. In other parts of the system, the system may be robust against fluctuations of large magnitude. In such a case, it may often suffice to confirm that the parameter values fall within the range of stability, instead of obtaining highly accurate figures. The point is that not all parts of the system need to be tuned with the same precision. For example, components for jet engines may have to be produced with high precision, but seat belts do not have to achieve the same precision as jet engine components. In future, the type and accuracy requirements for experiments may be determined by theoretical requirements.

The examples given so far have focused on the process of identification of network structure and parameters that enable simulation and analysis of biochemical networks under the simplified assumption that all materials are distributed homogeneously in the environment. Unfortunately, this is not the case in biological systems. There are subcellular structures and localization of transcription products that cause major diversion from a naive model. Multi-cellular systems require measurement of cell-cell contact, diffusion, cell lineage, gene expression during development, etc. For accurate simulation and analysis, these features have to be measured in a comprehensive, accurate, and systematic manner. We have not developed devices to obtain high-throughput measurements for any of these features. This is a serious issue that has to be addressed.

Next-generation Experimental Systems

To cope with increasing demands for comprehensive and accurate measurement, a set of new technologies and instruments needs to be developed that offers a higher level of automation and high-precision measurement.

First, dramatic progress in the level of automation of experimental procedures for routine experiments is required in order to keep up with increasing demands for modeling and system-level analysis. High-throughput experiments may turn into a labor-intensive nightmare unless the level of automation is drastically improved. Further automation of experimental procedures would greatly benefit the reliability of experiments, throughput, and total cost of the whole operation in the long run.

Second, cutting-edge technologies such as micro-fluid systems, nanotechnology and femto-chemistry may need to be introduced to design and build next-generation experimental devices. The use of such technologies will enable us to measure and observe the activities of genes and proteins in a way that is not possible today. It may also drastically improve the speed and accuracy of measurement for existing devices.

In those fields where there are obvious needs, such as sequencing and proteomics, the above goals are already pursued. Beyond the development of high-throughput sequencers using high-density capillary array electrophoresis, efforts are being made to develop integrated microfabricated devices that enable PCR and capillary electrophoresis in a single micro device (Lagally et al., 1999; Simpson et al., 1998). Such devices not only enable miniaturization and precision measurements, but will also significantly increase the level of automation.

In the developmental biology of *C. elegans*, identification of cell lineage is one of the major issues that needs to be accomplished to assist analysis of the gene regulatory network for differentiation. The first attempt to identify cell lineage was carried out entirely manually (Sulston et al., 1983; Sulston and Horvitz, 1977), and it took several years to identify the lineage of the wild type. Four-dimensional microscopy allowed us to collect multi-layer confocal images at a constant time interval, but lineage identification is not automatic. With the availability of exhaustive RNAi knockout for *C. elegans*, high-throughput cell lineage identification is essential to explore the utility of the exhaustive RNAi. Efforts are underway to fully automate cell lineage identification, as well as three-dimensional nuclei position data acquisition (Onami et al., 2001a), fully utilizing advanced image processing algorithms and massively parallel supercomputers. Such devices meet some of the criteria presented earlier, and provide comprehensive measurement of cell positions with high accuracy. With automation, high-throughput data acquisition can be expected. If the project succeeds, it can be used to automatically identify the cell lineage of all RNAi knockout for early embryogenesis. The technology may be augmented, but with major efforts, to automatically detect cell-cell contact, protein localization, etc.

Combined with whole mount *in situ* hybridization and possible future single-cell expression profiling, complete identification of the gene regulatory network for *C. elegans* may be possible in the near future.

SYSTEM STRUCTURE IDENTIFICATION

There are various system structures that need to be identified, such as the structural relationship among cells in the developmental process, detailed cell-cell contact configuration, membrane, intra-cellular structures, and gene regulatory networks. While each of these has significance in corresponding research in systems biology, this section focuses on how the structure of gene regulatory networks can be identified, primarily because it is a subject of growing interest due to the rapid uncovering of genomic information, and it is the control center of various cellular phenomena.

In order to understand a biological system, we must first identify the structure of the system. For example, to identify a gene regulatory network, one must identify all components of the network, the function of each component, interactions, and all associated parameters. All possible experimental data must be used to accomplish this non-trivial task. At the same time, inference results from existing experiments should enable the prediction of unknown genes and interactions, which can then be experimentally verified.

The difficulty is that such a network cannot be automatically inferred from experimental data based on some principles or universal rules, because biological systems evolve through stochastic processes and are not necessarily optimal. Also, there are multiple networks and parameter values that behave quite similar to the target network. One must identify the true network out of multiple candidates.

This process can be divided into two major tasks: (1) network structure identification, and (2) parameter identification.

Network Structure Identification

Several attempts have already been made to identify gene regulatory networks from experimental data. They can be classified into two approaches.

Bottom-up Approach

The bottom-up approach tries to construct a gene regulatory network based on the compilation of independent experimental data, mostly through literature searches and some specific experiments to obtain data of very specific aspects of the network of interest. Some of the early attempts of this approach are seen in the lambda phage decision circuit (McAdams and Shapiro, 1995), early embryogenesis of *Drosophila* (Reinitz et al., 1995; Hamahashi and Kitano, 1998; Kitano et al., 1997), leg formation (Kyoda and Kitano, 1999a), wing formation (Kyoda and Kitano, 1999b), eye formation on ommatidia clusters and R-cell differentiation (Morohashi and Kitano, 1998), and a reaction-diffusion based eye formation model (Ueda and Kitano, 1998). This approach is suitable when most of the genes and their regulatory relationship are relatively well understood. This approach is particularly suitable for the end-game scenario where most of the pieces are known and one is trying to find the last few pieces. In some cases, biochemical constants can be measured so that very precise simulation can be performed. When most parameters are available, the main purpose of the research is to build a precise simulation model which can be used to analyze the dynamic properties of the system by changing the parameters that cannot be done in the actual system, and to confirm that available knowledge generates simulation results that are consistent with available experimental data.

There are efforts to create databases that describe gene and metabolic pathways from the literature. KEGG (Kanehisa and Goto, 2000) and EcoCyc (Karp et al., 1999) are typical examples. Such databases are enormously useful for modeling and simulation, but they must be accurate and represented in such a way that simulation and analysis can be done smoothly.

There have been some preliminary attempts to predict unknown genes and their interactions (Morohashi and Kitano, 1998; Kyoda and Kitano, 1999a,b). These attempts manually searched possible unknown interactions to obtain simulation results consistent with experimental data, and did not perform exhaustive searches of all possible spaces of network structures.

Top-down Approach

The top-down approach tries to make use of high-throughput data using DNA microarray and other new measurement technologies. Already, there have been some attempts to infer groups of genes that have a tight relationship based on DNA microarray data using clustering techniques for the yeast cell cycle (Brown and Botstein, 1999; DeRisi et al., 1997; Spellman et al., 1998) and development of mouse central neural systems (D'haeseleer et al., 1999). Clustering methods are suitable for handling large-scale profile data, but do not directly deduce the network structures. Such methods only provide clusters of genes that are co-expressed in similar temporal patterns. Often, easy-to-understand visualization is required (Michaels et al., 1998).

Some heuristics must be imposed if we are to infer networks from such methods. Alternative methods are now being developed to directly infer network structures from expression profiles (Morohashi and Kitano, 1999; Liang et al., 1999) and extensive gene disruption data (Akutsu et al., 1999; Ideker et al., 2000). Most of the methods developed in the past translate expression data into binary values, so that the computing cost can be reduced. However, such methods seriously suffer from information loss in the binary translation process, and cannot obtain the accurate network structure. A method that can directly handle continuous-value expression data was proposed (Kyoda et al., 2000b; Onami et al., 2001b) and reported accurate performance without a serious increase in computational costs. An extension of this method seems to be very promising for any serious research on inference of gene regulatory networks.

Genetic programming has been applied to automatically reconstruct pathways and parameters that fit experimental data (Koza et al., 2001). The approach requires extensive computing power, and an example of such is the 1,000 CPU cluster Beowulf-class supercomputer, but the approach has the potential to be practical given the expected speed up of processor chips.

Such extensions include the development of a hybrid method that combines the bottom-up and the top-down approach. It is unlikely that no knowledge is available before applying any inference methods; in practical cases, it can be assumed that various genes and their interactions are partially understood, and that it is necessary to identify the rest of the network. By using knowledge that is sufficiently accurate, the possible space of network structures is significantly reduced.

One major problem is that such methods cannot directly infer possible modifications and translational control. Future research needs to address integration of the data of the expression profile, protein-protein interactions, and other experimental data.

Parameter Identification

It is important to identify only the structure of the network, but a set of parameters, because all computational results have to be matched and tested against actual experimental results. In addition, the identified network will be used for simulating a quantitative analysis of the system's response and behavioral profile.

In most cases, the parameter set has to be estimated based on experimental data. Various parameter optimization methods, such as genetic algorithms and simulated annealing, are used to find a set of parameters that can generate simulation results consistent with experimental data (Hamahashi and Kitano, 1999). In finding a parameter set, it must be noted that there may be multiple parameter sets which generate simulation results equally fitted to experimental data. An important feature of parameter optimization algorithms used for this purpose is the capability to find as many local minima (including a global minima) as possible, rather than finding single global minima. This needs to be combined with a method to indicate specific experiments to identify which one of such parameter sets is the correct parameter set.

There are several methods to find optimal parameter sets such as brute force exhaustive search, genetic algorithms, simulated annealing, etc. Most of them are computationally expensive, and have not been considered viable options in the past. But the situation has changed, and it will change in future, too.

Although it is important to accurately measure and estimate the genuine parameter values, in some cases parameters are not that critical. For example, it was shown through an extensive simulation that the segment polarity network in *Drosophila* exhibits a high level of robustness against parameter change (von Dassow et al., 2000). For certain networks that are essential for survival the networks need to be built robust against various changes in parameters to cope with genetic variations and external disturbances. For this kind of network, the essence is embedded into the structure of the network, rather than specific parameters of the network. This is particularly the case when feedback control is used to obtain robustness of the circuits, as seen in bacterial chemotaxis (Yi et al., 2000).

Thus, parameter estimation and measurement may need to be combined with theoretical analysis on sensitivity of certain parameters to maintain functionalities of the circuit.

SYSTEM BEHAVIOR ANALYSIS

Once we understand the structures of the system, research will focus on dynamic behaviors of the system. How does it adapt to changes in the environment, such as nutrition, and various stimuli? How does it maintain robustness against various potential damage to the system, such as DNA damage and mutation? How do specific circuits exhibit functions observed? To attain system-level understanding, it is essential to understand the mechanisms behind (1) the robustness and stability of the system, and (2) functionalities of the circuits.

It is not a trivial task to understand the behaviors of complex biological networks. Computer simulation and a set of theoretical analyses are essential to provide in-depth understanding on the mechanisms behind the circuits.

Simulation

Simulation of the behavior of gene and metabolism networks plays an important role in systems biology research, and there are several ongoing efforts on simulator development (Mendes and Kell, 1998; Tomita et al., 1999; Kyoda et al., 2000a; Nagasaki et al., 1999). Due to the complexity of the network behavior and large number of components involved, it is almost impossible to intuitively understand the behaviors of such networks. In addition, accurate simulation models are prerequisite for analyzing the dynamics of the system by changing the parameters and structure of the gene and metabolism networks. Although such analysis is necessary for understanding the dynamics, these operations are not possible with actual biological systems. Simulation is an essential tool not only for understanding the behavior, but also for the design process. In the design of complex engineering systems, various forms of simulation are used. It is unthinkable today that any serious engineering systems could be designed and built without simulation. VLSI design requires major design simulation, thus creating one of the major markets for supercomputers. Commercial aviation is another example. The Boeing 777 was designed based almost entirely on simulation and digital prefabrication. Once we enter that stage of designing and actively controlling biological systems, simulation will be the core of the design process.

For simulation to be a viable methodology for the study of biological systems, highly functional, accurate, and user-friendly simulator systems need to be developed. Simulators and associated software systems often require extensive computing power such that the system must run on highly parallel cluster machines, such as the Beowulf PC cluster (Okuno et al., 1999). Although there are some simulators, there is no system that sufficiently covers the needs of a broad range of biology research. Such simulators must be able to simulate gene expression, metabolism, and signal transduction for a single and multiple cells. It must be able to simulate both high concentration of proteins that can be described by differential equations, and low concentration of proteins that need to be handled by stochastic process simulation. Some efforts on simulating a stochastic process (McAdams and Arkin, 1998) and integrating it with high concentration level simulation are underway.

In some cases, the model requires not only gene regulatory networks and metabolic networks, but also high-level structures of chromosomes, such as heterochromatin structures. In the model of aging, some attempts are being made to model heterochromatin dynamics (Kitano and Imai, 1998; Imai and Kitano, 1998). Nevertheless, how to model such dynamics and how to estimate the structure from sparse data and the current level of understanding are major challenges.

The simulator needs to be coupled with parameter optimization tools, a hypothesis generator, and a group of analysis tools. Nevertheless, algorithms behind these software systems need to be designed precisely for biological research. One example that has already been mentioned is that the parameter optimizer needs to find as many local minima (including global minima) as possible, because there are multiple possible solutions of which only one is actually used. The assumption that the most optimal solution is used in an actual system does not hold true in biological systems. Most parameter optimization methods are designed to find the global optima for engineering design and problem solving. While existing algorithms provide a solid starting point, they must be modified to suit biological research. Similar arguments apply to other software tools, too.

A set of software systems needs to be developed and integrated to assist systems biology research. Such software includes:

- a database for storing experimental data,
- a cell and tissue simulator,
- parameter optimization software,
- bifurcation and systems analysis software,
- hypotheses generator and experiment planning advisor software, and
- data visualization software.

How these modules are related and used in an actual work flow is illustrated in Figure 1.1. While many independent efforts are being made on some of this software, so far only limited efforts have been made to create a common platform that integrates these modules. Recently, a group of researchers initiated a study to define a software platform for systems biology. Although various issues need to be addressed for such a software platform, the rest of this section describes some illustrative issues.

Efforts are being made to provide a common and versatile software platform for systems biology research. The Systems Biology Workbench project aims to provide a common middleware so that plug-in modules can be added to form a uniform software environment.

Beside the software module itself, the exchange of data and the interface between software modules is a critical issue in data-driven research tools. Systems Biology Mark-up Language (SBML) is a versatile and common open standard that enables the exchange of data and modeling information among a wide variety of software systems (Hucka et al., 2000, 2001). It is an extension of XML, and is expected to become the industrial and academic standard of the data and model exchange format.

Ultimately, a group of software tools needs to be used for disease modeling and simulation of organ growth and control; this requires a comprehensive and highly integrated simulation and analysis environment.

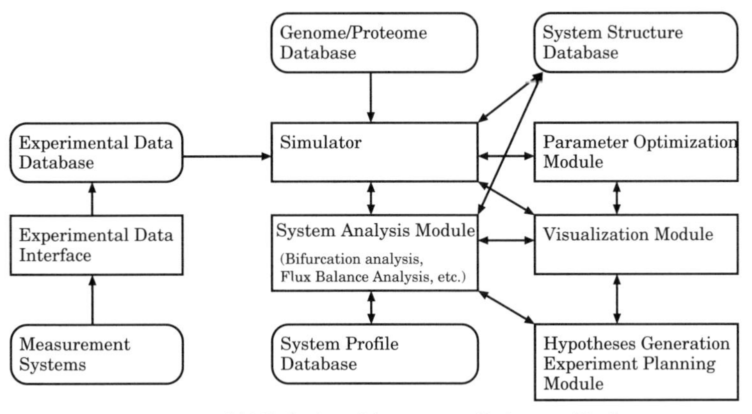

(A) Relationship among Software Tools

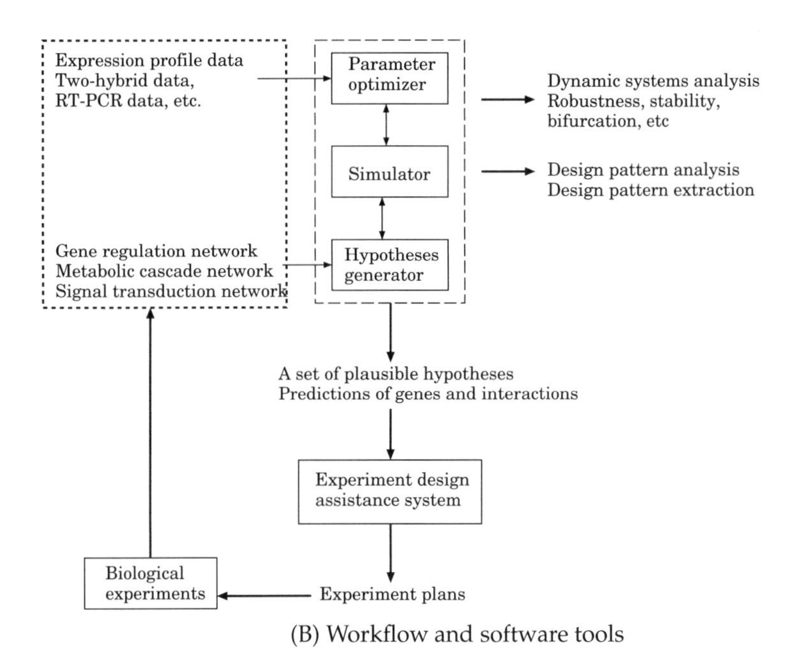

(B) Workflow and software tools

Figure 1.1 Software tools for systems biology and their workflow

Analysis Methods

There have been several attempts to understand the dynamic properties of systems using bifurcation analysis, metabolic control analysis, and sensitivity analysis. For example, bifurcation analysis has been used to understand the *Xenopus* cell cycle (Borisuk and Tyson, 1998). The analysis creates a phase portrait based on a set of equations describing the essential process of the *Xenopus* cell cycle. A phase portrait illustrates in which operation point the system is acting, and how it changes behavior if some of the system parameters are varied. By looking at the landscape of the phase portrait, a crude analysis of the robustness of the system can be made.

A group of analysis methods such as flux balance analysis (FBA) (Varma and Palsson, 1994; Edward and Palsson, 1999) and metabolic control analysis (MCA) (Kacser and Burns, 1973; Heinrich and Rapoport, 1974; Fell, 1996) provides a useful method to understand system-level behaviors of metabolic circuits under various environments and internal disruptions. It has been demonstrated that such an analysis method can provide knowledge on the capabilities of metabolic pathways that are consistent with experimental data (Edward et al., 2001). While such methods are currently aiming at analysis of the steady-state behaviors with linear approximation, extention to dynamic and nonlinear analysis would certainly provide a powerful tool for system-level analysis of metabolic circuits.

Several other analysis methods have already been developed for complex engineering systems, particularly in the area of control dynamic systems. One of the major challenges is to describe biological systems in the language of control theory, so that we can abstract essential parts of the system within the common language of biology and engineering.

ROBUSTNESS OF BIOLOGICAL SYSTEMS

Robustness is one of the essential features of biological systems. Understanding the mechanism behind robustness is particularly important because it provides in-depth understanding on how the system maintains its functional properties against various disturbances. Specifically, we should be able to understand how organisms respond to (1) changes in environment (deprived nutrition level, chemical attractant, exposure to various chemical agents that bind to receptors, temperature) and (2) internal failures (DNA damage, genetic malfunctions in metabolic pathways). Obviously, it is critically important to understand the intrinsic functions of the system, if we are eventually to find cures for diseases.

Lessons from Complex Engineering Systems

There are interesting analogies between biological systems and engineering systems. Both systems are designed incrementally through some sort of evolutionary processes, and are generally suboptimal for the given task. They also exhibit increased complexity to attain a higher level of robustness and stability.

Consider an airplane as an example. If the atmospheric air flow is stable and the airplane does not need to change course, altitude, or weight balance, and does not need to take off and land, the airplane can be built using only a handful of components. The first airplane built by the Wright brothers consisted of only a hundred or so components. The modern jet, such as the Boeing 747, consists of millions of components. One of the major reasons for the increased complexity is to improve stability and robustness. Is this also the case in biological systems?

Mycoplasma is the smallest self-sustaining organism and has only about 400 genes. It can only live under specific conditions, and is very vulnerable to environmental fluctuations. *E. coli*, on the other hand, has over 4,000 genes and can live under varying environments. As *E. coli* evolved it acquired genetic and biochemical circuits for various stress responses and basic behavioral strategies such as chemotaxis (Alon et al., 1999; Barkai and Leibler, 1997). These response circuits form a class of negative feedback loop. Similar mechanisms exist even in eukaryotic cells[2].

A crude speculation is that further increases in complexity in multicellular systems toward *homo sapiens* may add functionalities that can cope with various situations in their respective ecological niche.

In engineering systems, robustness and stability are achieved by the use of (1) system control, (2) redundancy, (3) modular design, and (4) structural stability. The hypothesis is that the use of such an approach is an intrinsic feature of complex systems, be they artificial or natural.

System Control: Various control schemes used in complex engineering systems are also found in various aspects of biological systems. Feedforward control and feedback control are two major control schemes, both of which are found almost ubiquitously in biological systems. Feedforward control is an open-loop control in which a set of predefined reaction sequences is triggered by a certain stimulus. Feedback is a sophisticated control system that closes the loop of the signal circuits to attain the desired control of the system. A negative feedback system detects the difference between desired output and actual output and compensates for such difference by modulating the input. While there are feedforward control methods, feedback control is more sophisticated and ensures proper control of the system and it can be used with feedforward control. It is one of the most widely used methods in engineering systems to increase the stability and robustness of the system.

Redundancy: Redundancy is a widely used method to improve the system's robustness against damage to its components by using multiple pathways to accomplish the function. Duplicated genes and genes with similar functions are basic examples of redundancy. There is also circuit-level redundancy, such as multiple pathways of signal transduction and metabolic circuits that can be functionally complementary under different conditions.

Modular Design: Modular design prevents damage from spreading limitlessly, and also improves ease of evolutionary upgrading of some of the

2 Discussion of similarity between complexity of engineering and biological systems as described in this section was first made, as far as the author is aware, by John Doyle at Caltech.

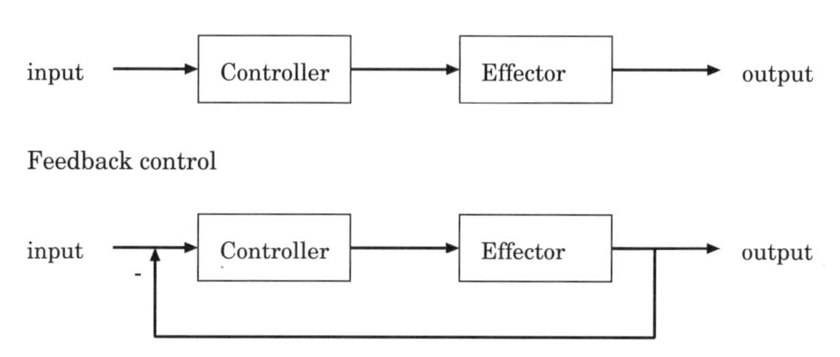

Figure 1.2 Feedforward control and feedback control

components. At the same time, a multi-functional module can help overcome system failure in a critical part by using modules in other less critical parts. Cellular systems are typical examples of modular systems.

Structural Stability: Some gene regulatory circuits are built to be stable for a broad range of parameter variations and genetic polymorphisms. Such circuits often incorporate multiple attractors, each of which corresponds to functional state of the circuit; thus its functions are maintained against change in parameters and genetic polymorphisms.

It is not clear whether such engineering wisdom is also the case in biological systems. However, the hypothesis is that such features are somewhat universal in all complex systems. It is conceivable that there are certain differences due to the nature of the system it is built upon, as well as the difference between engineering systems that are designed to exhibit certain functions and natural systems that have reproduction as a single goal where all functions are only evaluated in an integrated effect. Nevertheless, it is worth investigating the univerality of principles. And, if there are differences, what are they?

The rest of the section focuses on how three principles of robustness exist also in biological systems. Of course, not all biological systems are robust, and it is important to know which parts of the systems are not robust and why. However, for this particular chapter, we will focus on robustness of biological systems, because it is one of the most interesting issues that we wish to understand.

Control

The use of explicit control scheme is an effective approach to improving robustness. Feedforward control and feedback control are two major methods of system control (Figure 1.2).

Feedforward control is an open-loop control in which a sequence of predefined actions is triggered by a certain stimulus. This control method is the simplest method that works when possible situations and countermeasures are highly predictable.

Feedback control, such as negative feedback, is a sophisticated control method widely used in engineering. It feeds back the sign-inverted error between the desired value and the actual value to the input, then the input signal is modulated proportional to the amount of error. In its basic form, it acts to minimize the output error value.

Feedback plays a major role in various aspects of biological processes, such as *E. coli* chemotaxis and heat shock response, circadian rhythms, cell cycle, and various aspects of development.

The most typical example is the integral feedback circuits involved in bacterial chemotaxis. Bacteria demonstrates robust adaptation to a broad range of chemical attractant concentrations, and so can always sense changes in chemical concentration to determine its behavior. This is accomplished by a circuit that involves a closed-loop feedback circuit (Alon et al., 1999; Barkai and Leibler, 1997). As shown in Figure 1.3, ligands that are involved in chemotaxis bind to a specific receptor MCP that forms a stable complex with CheA and CheW. CheA phosphorylates CheB and CheY. Phosphorylated CheB demethylates the MCP complex, and phosphorylated CheY triggers tumbling behavior. It was shown through experiments and simulation studies that this forms a feedback circuit which enables adaptation to changes in ligand concentration. Specifically, for any sudden change in the ligand concentration, the average activity level that is characterized by the tumbling frequency quickly converges to the steady-state value. This means that the system only detects acute changes of the ligand concentration that can be exploited to determine tumbling frequency, but is insensitive to the absolute value of ligand concentration. Therefore, the system can detect and control its behavior to move to a high attractant concentration area in the field regardless of the absolute concentration level without saturating its sensory system. Detailed analysis revealed that this circuit functions as an integral feedback (Yi et al., 2000) — the most typical automatic control strategy.

In bacteria, there are many examples of sophisticated control embedded in the system. The circuit that copes with heat shock, for example, is a beautiful example of the combined use of feedforward control and feedback control (Figure 1.4). Upon heat shock, proteins in *E. coli* can no longer maintain their normal folding structures. The goal of the control system is to repair misfolding proteins by activating a heat shock protein (hsp), or to dissociate misfolding proteins by protease. As soon as heat shock is imposed, a quick translational modulation facilitates the production of σ^{32} factor by affecting the three-dimensional structure of rpoH mRNA that encodes σ^{32}. This leads to the formation of σ^{32}-RNAP holo-enzyme that activates hsp that repair misfolded proteins. This process is feedforward control that pre-encodes the relationship between heat shock and proper course of reactions. In this process, there is no detection of misfolded pro-

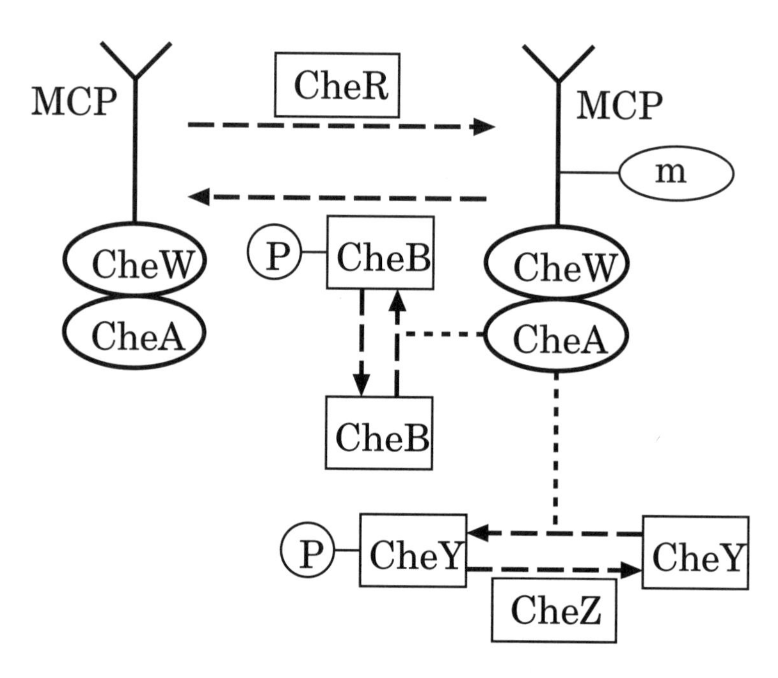

Figure 1.3 Bacterial chemotaxis related feedback loop

teins to adjust the translational activity of σ^{32}. Independently, DnaK and DnaJ detect misfolded proteins and release σ^{32} factor, that has been bound with DnaK and DnaJ. Free σ^{32} activates transcription of hsp, so that misfolded proteins are repaired. This process is negative feedback control, because the level of misfolded proteins is monitored and it controls the activity of σ^{32} factor.

Another example demonstrating the critical role of the feedback system is seen in growth control of human cells. Growth control is one of the most critical parts of cellular functions. The feedback circuit involved in p53 presents a clear example of how feedback is used (Figure 1.5). When DNA is damaged, DNA-dependent kinase DNA-PK is activated. Also, *ATM* is phosphorylated, which makes *ATM* itself in an active state and promotes phosphorylation of the specific locus of the p53 protein. When this locus is phosphorylated, p53 no longer forms a complex with MDM2, and escapes from dissociation. The phosphorylation locus depends on what kind of stress is imposed on DNA. Under a certain stress, phosphorylation takes place at the Ser15 site of p53, and promotes transcription of p21 that eventually causes G1 arrest. In other cases, it promotes activation of apoptosis inducing genes, such as pig-3, and results in apoptosis. For those cells that entered G1 arrest, DNA-PK and ATM activity are lost as soon as DNA is repaired. The loss of DNA-PK and ATM activity decreases phosphorylation of p53, so p53 will bind with MDM2 and dissolve.

Without phosphorylation, the p53 protein promotes mdm-2 transcription. It is interesting to know that mdm-2 protein forms a complex to deac-

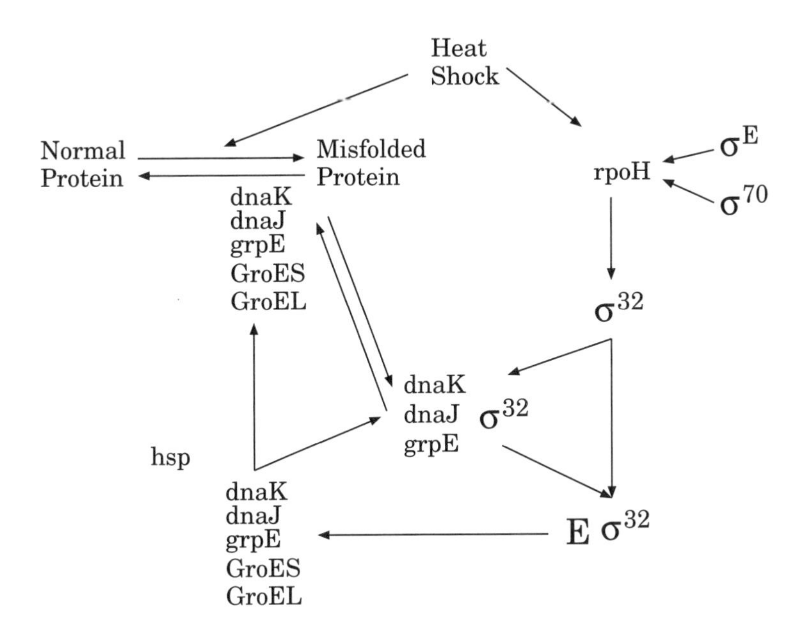

Figure 1.4 Heat shock response with feedforward and feedback control

tivate the p53 protein. This is another negative feedback loop embedded in this system.

Redundancy

Redundancy also plays an important role in attaining robustness of the system, and is critical for coping with accidental damage to components of the system. For example, the four independent hydraulic control systems in a Boeing 747 render the systems functionally normally even if one or two of them are damaged. In aircraft, control systems and engines are designed to have a high level of redundancy. In a cellular system, signal transduction and cell cycle are equivalent to control systems and engines.

A typical signal transduction pathway is the MAP kinase cascade. The MAP kinase cascade involves extensive cross talk among collateral pathways. Even if one of these pathways is disabled due to mutation or other causes, the function of the MAP kinase pathway can be maintained because other pathways still transduce the signal (Figure 1.6).

Cell cycle is the essential process of cellular activity. For example, in the yeast cell cycle, the Cln and Clb families play a dominant role in the progress of the cell cycle. They bind with Cdc28 kinase to form Cdk complex. Cln is redundant because knock-out of up to two of three Cln (Cln1, Cln2, Cln3) does not affect the cell cycle; all three Cln have to be knocked out to stop the cell cycle. Six Clb have very similar features, and may have originated in gene duplication. No single loss-of-function mutant of any of the six Clb affects growth of the yeast cell. The double mutants of CLB1

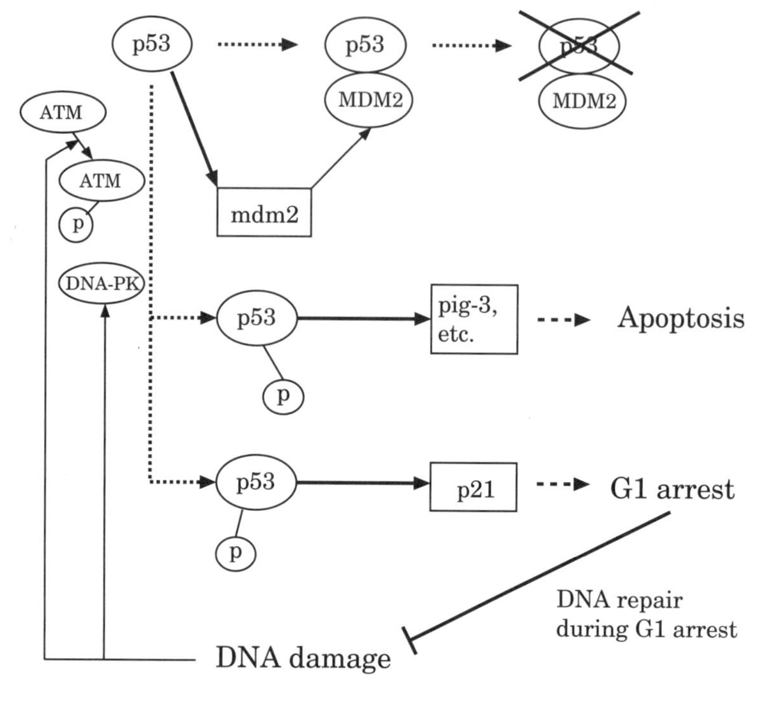

Figure 1.5 p53 related feedback loop

and CLB2, as well as CLB2 and CLB3s are lethal, but other double mutant combinations do not affect phenotype. It is reasonable that the basic mechanism of the cell cycle has evolved to be redundant, thus robust against various perturbations.

Redundancy can be exploited to cope with uncertainty involved in stochastic processes. McAdams and Arkin argued that duplication of genes and the existence of homologous genes improve reliability so that transcription of genes can be carried out even when only a small number of transcription factors are available (McAdams and Arkin, 1999). The use of a positive feedback loop to autoregulate a gene to maintain its own expression level is an effective means of ensuring the trigger is not lost in the noise.

Although its functional implication has not been sufficiently investigated, an analysis of MAP kinase cascade revealed that it utilizes non-linear properties intrinsic in each step of the cascade and positive feedback to constitute a stable all-or-none switch (Ferrell and Machleder, 1998).

In the broader sense, the existence of metabolic pathways that can alternatively function to sustain cellular growth with changing environment can be viewed as redundancy. Bacteria is known to switch metabolic pathways if deprived of one type of nutrition, and to use other types of nutrition that are available. Theoretical analysis combined with experimental data indicate that different pathways are used to attain essentially the

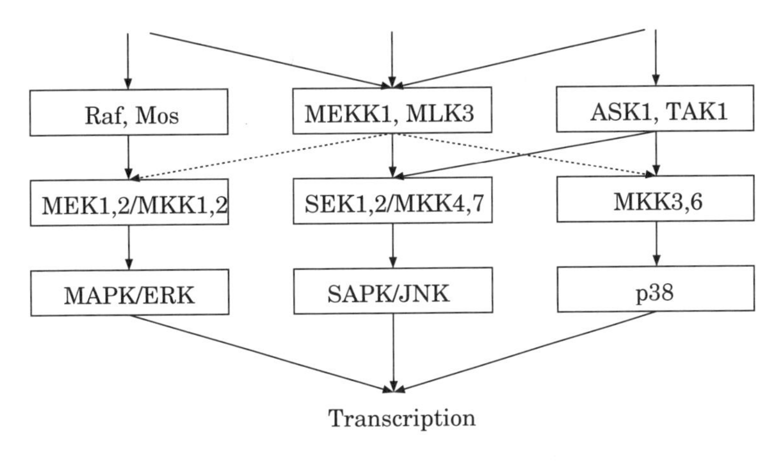

Figure 1.6 Redundancy in MAP kinase cascade

same objective function (Edward et al., 2001).

Once we understand the stability and robustness of the system, we should be able to understand how to control and transform cells. We will then be ready to address such questions as how to transform cells that are malfunctioning into normal cells, how to predict disease risk, and how to preemptively treat potential diseases.

Modular Design

Modular design is a critical aspect of the robustness: it ensures that damage in one part of the system does not spread to the entire system. It may also ensure efficient reconfiguration throughout the evolutionary process to acquire new features.

The cellular structure of the multicellular organism is a clear example. It physically partitions the structure so that the entire system does not collapse due to local damage.

Gene regulatory circuits are considered to entail a certain level of modularity. Even if part of the circuit is disrupted due to mutation or injection of chemicals, it does not necessary affect other parts of the circuit. For example, mutation in p53 may destroy the cell cycle check point system that leads to cancer. However, it does not destroy metabolic pathways, so the cells continue to proliferate. How and why such modularity is maintained is not well understood at present.

Modularity reflects hierarchical organization of the system that can be viewed as follows:

Component: An elementary unit of the system. In electronics, transistors, capacitors, and resistors are components. In biological systems, genes and proteins, which are transcriptional products, are components.

Device: A minimum unit of the functional assembly. NAND gates and flip-flops are examples of devices[3]. Transcription complexes and replication complexes are examples of devices. Some signal transduction circuits may be considered as devices.

Module: A large cluster of devices. CPU, memory, and amplifiers are modules. In biological systems, organella and gene regulatory circuits for the cell cycle are examples of modules.

System: A top-level assembly of modules. Depending on the viewpoint, a cell or entire animal can be considered as a system.

In engineering wisdom, each low-level module should be sufficiently self-contained and encapsulated so that changes in higher-level structure do not affect internal dynamics of the lower-level module. Whether is this also the case for biological systems and how it can be accomplished are of major interest from a system perspective.

Structural Stability

Some circuits may, after various disturbances to the state of the system, resume as one of multiple attractors (points or periodic). Often, feedback loops play a major role in making this possible. However, feedback does not explicitly control the state of the circuit in tracking or adapting to stimuli. Rather, dynamics of the circuit exhibit certain functions that are used in the larger sub-systems.

The most well understood example is seen in one of the simplest organisms, lambda phage (McAdams and Shapiro, 1995). Lambda phage exploits the feedback mechanism to stabilize the committed state and to enable switching of its pathways. When lambda phage infects *E. coli*, it chooses one of two pathways: lysogeny and lysis. While a stochastic process is involved in the early stage of commitment, two positive and negative feedback loops involving CI and Cro play a critical role in stable maintenance of the committed decision. In this case, whether to maintain feedback or not is determined by the amount of activator binding to the O_R region, and the activator itself cuts off feedback if the amount exceeds a certain level. This is an interesting molecular switch that is not found elsewhere. Overall, the concentration mechanism of Cro is maintained at a certain level using positive feedback and negative feedback. It was reported that the fundamental properties of the lambda phage switch circuit are not affected even if the sequence of O_R binding sites is altered (Little et al., 1999). This indicates that properties of the lambda phage decision circuit are intrinsic to the multiple feedback circuit, not specific parametric features of the elements, such as binding sites.

Relative independence from specific parameters is an important fea-

3 In electronics, "device" means transistors and other materials mentioned in "components." NAND gates and flip-flops are recognized as minimum units of the circuit.

ture of a robust system. Recent computational studies report that circuits that are robust against a broad range of parameter variations are found in *Xenopus* cell cycle (Morohashi et al., unpublished) and body segment formation (von Dassow et al., 2000). Using the simulation of parasegment formation of *Drosophila*, it was found that some parameters in the circuit accountable for pattern formation are tolerant to major parameter variations. This strongly suggests that the structure of the circuit that is dominantly responsible for pattern formation rather than specific parameter values (von Dassow et al., 2000).

Such circuit features of structural stability also play important roles in development. A recent review article (Freeman, 2000) elucidates some interesting cases of feedback circuits that play a dominant role in the development process. Such cases include temporal arrangement of signaling in the JAK/STAT signaling pathway, pattern formation in *Drosophila* involving Ubx and Dpp, maintenance of patterns of expression for sonic hedgehog (Shh) that forms ZPA and Fgf, forming AER in limb development, etc. In these examples, structure of circuits play the dominant role rather than specific set of parameters.

THE SYSTEOME PROJECT

In order to promote scientific research of systems biology, it is critically important to create a comprehensive data resource that describes systems' features, as does the human genome project. This is an enormous challenge, and it requires significant efforts far beyond the capability of any single research group. Therefore, the author proposes "The Systeome Project" as a grand challenge in the area of systems biology.

Systeome is an assembly of system profiles for all genetic variations and environmental stimuli responses. A system profile comprises a set of information on the properties of the system that includes the structure of the system and its behaviors, analysis results such as phase portfolio, and bifurcation diagrams. The structure of the system includes the structure of gene and metabolic networks and its associated constants, physical structures and their properties.

Systeome is different from a simple cascade map, because it assumes active and dynamic simulations and profiling of various system statuses, not a static entity. The author suggests that the project be established for comprehensive efforts for profiling the Systeome of human, mouse, *Drosophila, C. elegans*, and yeast.

The goal of the Human Systeome Project is defined as "To complete a detailed and comprehensive simulation model of the human cell at an estimated error margin of 20 percent by the year 2020, and to finish identifying the system profile for all genetic variations, drug responses, and environmental stimuli by the year 2030."

Undoubtedly, this is an ambitious project, and requires several mile-

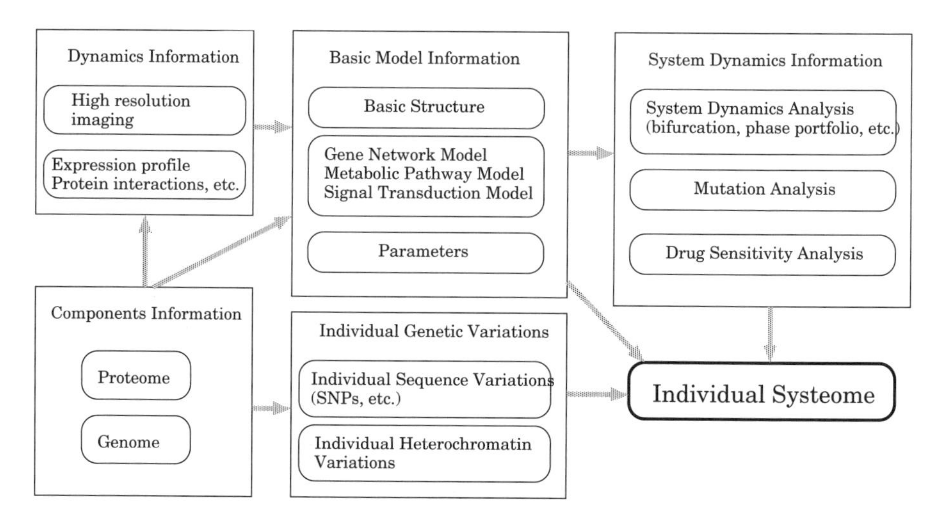

Figure 1.7 Genome, Proteome, and Systeome

stones and pilot projects leading to the final goal. Initial pilot projects can using yeast and *C. elegans* be set with a time frame of five or seven years after full-scale budget approval. The Human Systeome Project shall be commenced concurrently with such pilot projects.

The impact of this project will be far-reaching. It will be a standard asset for biological research as well as providing fundamental diagnostics and prediction for a wide range of medical practices.

The Systeome Project is expected to contribute to system-level understanding of life by providing exhaustive knowledge of system structures, dynamics, and their sensitivities against genetic variations and environmental stimuli. By using the system profile, it is expected that more precise medical diagnosis and treatment can be accomplished due to quantitative understanding of the metabolic state of the system. For example, a list of all possible feedback loops and their sensitivities, gain, and time delay should be obtained, to be used for drug design and clinical applications. The behaviors of feedback systems are often counterintuitive and often eliminate or compensate the effects of external stimuli. Understanding of complex circuit dynamics such as these will contribute to accurate prediction of the effects of medical treatments.

The Systeome Project should maintain close links with genome and Proteome data, particularly with various individual genetic variations, including single nucleotide polymorphisms (SNPs). SNPs are a typical example of an attempt to understand the relationship between genetic variations and clinical observations.

It is inevitable that in some cases the effects of SNPs are masked by a mechanism that compensates such variations. In this case, corresponding SNPs do not seem to affect the behavior of the cell. However, if such a compensation mechanism is disrupted by SNPs in a locus that constitutes the compensation mechanism, the effects of SNPs will show up directly in the cell's behavior. In such a case, it will be observed that for certain groups of cells, SNPs affect phenotype, but for other groups SNPs do not seem to affect phenotype.

While SNPs provide certain information on individual variations at the genetic level, they do not provide the quantitative status of mRNA and proteins. Many biological phenomena have a certain quantitative sensitivity. The cell cycle, for example, is expected to take place when cyclin synthesis and degradation rate are within a certain range. SNPs and other existing genetic analysis cannot provide insights into quantitative aspects of such phenomena.

Scientifically, a detailed understanding of circuits and their dynamics will contribute to a deeper understanding of the biological systems, as already discussed elsewhere.

Identification of metabolic and signal transduction circuits in various model systems provides an interesting opportunity to compare evolutionary conserved genetic information not only at the gene level, but also at the circuit level.

Evolutionary conserved circuits will be an important concept that may be widely used in the study of gene and metabolic network behaviors. Several circuits that may be found in yeast and *C. elegans* may be used also in mouse and human, similar to the idea of homologue genes.

Some of the feedback circuits, for example, may be so essential that they have been conserved through the course of evolution. At the same time, a certain circuit may be duplicated and a revised version is used for other parts of the system. As the Systeome Project progresses in various model systems, such comparative studies and homology searches at the circuit level will become possible.

Many scientific opportunities will open up once the Systeome Project has commenced and its data is made available for scientific research.

The Systeome Project will be a major commitment. However, it is indispensable for promoting systems biology as quickly as possible and for contributing to a better understanding of living systems and for medical practice. The Systeome Project involves the major engineering project of developing the measurement and software platform. The best way to proceed with this project is to initiate it as an international joint project on a scale comparable to the human genome project.

IMPACTS OF SYSTEMS BIOLOGY

Combined with the Systeome Project and other efforts in medical application of genomics, systems biology may have major impacts on medical research and practice. In-depth knowledge of the dynamical state of cells and development of high-performance measurement systems will drastically change medical practice.

First, the fast and precise measurement of an individual systeome will enable us to make precise assessment and simulation of disease risk, as well as detailed planning of countermeasures. Establishment of "preemptive molecular medicine" is one of the major applications of systems biology research. This means that patient models, or disease models, can be grounded on the cellular model, instead of being an empirical phenomenological model.

Second, drug design and treatment procedure may change to reflect the precise system dynamics of each patient. Rather than rely on a single drug, there many be increasing use of system drugs, a group of drugs that cooperatively act to control the metabolic state of malfunctioning cells. The point of such a treatment is to minimize side-effects, while maintaining maximum efficacy in disease treatment. By specifically identifying a series of effector points of chemical agents, we may be able to control cell status much more effectively than current medical practice.

Third, system-level understanding, especially simulation, control, and design capability, may lead to a totally new method of organ cloning. Just like engineers perform digital pre-assembly, we may be able to digitally pregrow organs for transplant. There will be a special incubation system that can monitor and control a growing organ inside the incubator. Currently, regenerative medicine is now being practiced, but it is limited to re-generation of relatively simple tissue systems such as skin. For growing more complex organs such as the heart and kidney, sophisticated growth monitoring and control are required. This is "closed-loop manufacturing," where the growth process is monitored and data is fed back to control the biochemical status of the incubation system to guide the organ growth to the desired shape.

There will be many more medical applications. The Systeome Project is perhaps the best way to accelerate progress in the technology of system-level biology.

CONCLUSION

Systems biology is a new and emerging field in biology that aims at system-level understanding of biological systems. System-level understanding requires a range of new analysis techniques, measurement technologies, experimental methods, software tools, and new concepts for looking at biological systems. The work has just begun and there is a long way to go before we arrive at a deep understanding of biological systems. Nevertheless, the author believes that systems biology will be the dominant paradigm in biology, and many medical applications as well as scientific discoveries are expected.

ACKNOWLEDGEMENTS

The author would like to thank members of the Kitano Symbiotic Systems Project (Shuichi Onami, Shugo Hamahashi, Koji Kyoda, Mineo Morohashi, John Doyle, Mel Simon, Hamid Bolouri, Tau-Mu Yi, Mark Borisuk, Michael Hucka, Andrew Finny, Herbert Sauro, Yoshi Kubota) for fruitful discussions that helped me to form the idea in this chapter. In particular, John and Mel have always been strong supporters of systems biology and sources of inspiration. Shin-ichirou Imai has always been a great collaborator, and it was he who guided me to the area of biology, while I was still absorbed in computer science and robotics. Mario Tokoro and Toshi Doi allowed me to work on biology despite its tenuous link with the ongoing business of Sony.

References

Akutsu, T., Miyano, S., and Kuhara, S. (1999). Identification of genetic networks from a small number of gene expression patterns under the Boolean network model. *Proc. Pacific Symposium on Biocomputing '99* pp.17–28.

Alon, U., Surette, M.G., Barkai, N., and Leibler, S. (1999). Robustness in bacterial chemotaxis. *Nature* 397:168–171.

Barkai, N. and Leibler, S. (1997). Robustness in simple biochemical networks. *Nature* 387:913–917.

Borisuk, M. and Tyson, J. (1998). Bifurcation analysis of a model of mitotic control in frog eggs. *Journal of Theoretical Biology* 195:69–85.

Brown, P.O. and Botstein, D. (1999). Exploring the new world of the genome with DNA microarrays. *Nature Genetics* 21:33–37.

Cannon, W.B., (1933). *The wisdom of the body,* Norton, New York.

The C. elegans Sequencing Consortium. (1998). Genome sequence of the nematode *C. elegans*: A platform for investigating biology. *Science* 282:2012–2018.

DeRisi, J.L., Lyer, V.R., and Brown, P.O. (1997). Exploring the metabolic and genetic control of gene expression on a genomic scale. *Science* 278:680–686.

D'haeseleer, P., Wen, X., Fuhrman, S., and Somogyi, R. (1999). Linear modeling of mRNA expression levels during CNS development and injury. *Proc. Pacific Symposium on Biocomputing '99* pp.41–52.

Edward, J.S. and Palsson, B.O. (1999). Systems properties of the Haemophilus influenzae Rd metabolic genotype. *Journal of Biological Chemistry* 274:17410–17416.

Edward, J.S., Ibarra, R., and Palsson, B.O. (2001). *In silico* predictions of *Escherichia coli* metabolic capabilities are consistent with experimental data. *Nature Biotechnology* 19(2):125–130.

Fell, D.A. (1996). *Understanding the control of metabolism,* Portland Press, London.

Ferrell, J. and Machleder, E. (1998). The biochemical basis of an all-or-none cell fate switch in *Xenopus* ooctytes. *Science* 280:895–898.

Freeman, M. (2000). Feedback control in intercellular signalling in development. *Nature* 408:313–319.

Hamahashi, S. and Kitano, H. (1998). Simulation of fly embryogenesis. *Proc. the 6th International Conference on Artificial Life* pp.151–160.

Hamahashi, S. and Kitano, H. (1999). Parameter optimization in hierarchical structure. *Proc. the 5th European Conference on Artificial Life* pp.467–471.

Heinrich, R. and Rapoport, T.A. (1974). A linear steady-state treatment of enzymatic chains. *Eur. J. Biochem.* 42:89–95.

Hucka, M., Sauro, H., Finney, A., and Bolouri, H. (2000). An XML-based model description language for systems biology simulations. *Working Draft, ERATO Kitano Project – CALTECH Group.*

Hucka, M., Finney, A., Sauro, H., Bolouri, H., Doyle, J., and Kitano, H. (2001). The ERATO Systems Biology Workbench: An integrated environment for multiscale and multitheoretic simulations in systems biology. *Foundations of Systems Biology,* The MIT Press, Cambridge.

Ideker, T., Thorsson, V., and Karp, R. (2000). Discovery of regulatory interactions through perturbation: inference and experimental design. *Proc. Pacific Symposium on Biocomputing 2000* pp.302–313.

Imai, S. and Kitano, H. (1998). Heterochromatin island and their dynamic reorganization: A hypothesis for three distinctive features of cellular aging. *Experimental Gerontology* 33(6):555–570.

Ito, T., Tashiro, K., Muta, S., Ozawa, R., Chiba, T., Nishizawa, M., Yamamoto, K., Kuhara, S., and Sakaki, Y. (2000). Toward a protein-protein interaction map of the budding yeast: A comprehensive system to examine two-hybrid interactions in all possible combinations between the yeast proteins. *Proc. Natl. Acad. Sci. USA* 97(3):1143–1147.

Kacser, H. and Burns, J. A. (1973). The control of flux. *Symp. Soc. Exp. Biol.* 27:65–104.

Kanehisa, M., and Goto, S. (2000). KEGG: Kyoto encyclopedia of gene and genomes. *Nucleic Acids Res.* 28:29–34.

Karp, P., Paley, M., Pellegrini-Toole, A., Krummenacker, M. (1999). EcoCyc: Electronic encycropedia of *E. coli* genes and metabolism. *Nucleic Acids Res.* 27(1):55.

Kitano, H. (2000). Perspectives on systems biology. *New Generation Computing* 18(3):199–216.

Kitano, H. and Imai, S. (1998). The two-process model of cellular aging. *Experimental Gerontology* 33(5):393–419.

Kitano, H., Hamahashi, S., Takao, K., and Imai, S. (1997). Virtual biology laboratory: A new approach of computational biology. *Proc. the 4th European Conference on Artificial Life* pp.274–283.

Kitano, H., Hamahashi, S., and Luke, S. (1998). The Perfect C. elegans Project: An initial report. *Artificial Life* 4:141–156.

Kondo, S. and Asai, R. (1995). A reaction-diffusion wave on the skin of the marine angelfish *Pomacanthus. Nature* 376:765–768.

Koza, J., Mydlowec, W., Lanza, G., Yu, J., and Keane, A. (2001). Automated reverse engineering of metabolic pathways from observed data by means of genetic programming. *Foundations of Systems Biology,* The MIT Press, Cambridge.

Kyoda, K. and Kitano, H. (1999). Simulation of genetic interaction for *Drosophila* leg formation. *Proc. Pacific Symposium on Biocomputing '99* pp.77–89.

Kyoda, K. and Kitano, H. (1999). A model of axis determination for the *Drosophila* wing disc. *Proc. the 5th European Conference on Artificial Life* pp.472–476.

Kyoda, K., Muraki, M., and Kitano, H. (2000). Construction of a generalized simulator for multi-cellular organisms and its application to SMAD signal transduction. *Proc. Pacific Symposium on Biocomputing 2000* pp.314–325.

Kyoda, K., Morohashi, M., Onami, S. and Kitano, H. (2000). A gene network inference method from continuous-value gene expression data of wild-type and mutants. *Genome Informatics* 11:196–204.

Lagally, E.T., Medintz, I., and Mathies, R.A. (2001). Single-molecule DNA amplication and analysis in an integrated microfluidic device. *Anal. Chem.* 73:565–570.

Liang, S., Fuhrman, S., and Somogyi, R. (1999). REVEAL, a general reverse engineering algorithm for inference of genetic network architectures. *Proc. Pacific Symposium on Biocomputing '99* pp.18–29.

Little, J.W., Shepley, D.P., and Wert, D.W. (1999). Robustness of a gene regulatory circuit. *EMBO J.* 18(15):4299–4307.

McAdams, H.H. and Arkin, A. (1999). It's a noisy business! Genetic regulation at the nanomolar scale. *Trends in Genetics* 15(2):65–69.

McAdams, H.H. and Arkin, A. (1998). Simulation of prokaryotic genetic circuits. *Annu. Rev. Biophys. Biomol. Struct.* 27:199–224.

McAdams, H. and Shapiro, L. (1995). Circuit Simulation of genetic networks. *Science* 269:650–656.

Mendes, P. and Kell, D.B. (1998). Non-linear optimization of biochemical pathways: Applications to metabolic engineering and parameter estimation. *Bioinformatics* 14(10):869–883.

Michaels, G.S., Carr, D.B., Askenazi, M., Fuhrman, S., Wen, X., and Somogyi, R. (1998). Cluster analysis and data visualization of large-scale gene expression data. *Proc. Pacific Symposium on Biocomputing'98* pp.42–53.

Morohashi, M. and Kitano, H. (1998). A method for reconstructing genetic regulatory network for *Drosophila* eye formation. *Proc. the 6th International Conference on Artificial Life* pp.72–80.

Morohashi, M. and Kitano, H. (1999). Identifying gene regulatory networks from time series expression data by *in silico* sampling and screening. *Proc. the 5th European Conference on Artificial Life* pp.477–486.

Morohashi, M., Winn, A.E., Borisuk, M.T., Bolouri, H., Doyle, J., and Kitano, H. Robustness as a measure of plausibility in models of biochemical networks. Unpublished.

Nagasaki, M., Onami, S., Miyano, S., and Kitano, H. (1999). Bio-Calculus: Its concept and molecular interaction. *Genome Informatics* 10:133–143.

Okuno, G.H., Kyoda, K., Morohashi, M., and Kitano, H. (1999). An initial assessment of ERATO-1 Beowulf-class cluster. *Proc. International Workshop on Parallel and Distributed Computing for Symbolic and Irregular Applications.*

Onami, S., Hamahashi, S., Nagasaki, M., Miyano, S., and Kitano, H. (2001). Automatic acquisition of cell lineage through 4D microscopy and analysis of early *C. elegans* embryogenesis. *Foundations of Systems Biology,* The MIT Press, Cambridge.

Onami, S., Kyoda, K.M., Morohashi, M., and Kitano, H. (2001). The DBRF method for inferring a gene network from large-scale steady-state gene expression data mutants. *Foundations of Systems Biology,* The MIT Press, Cambridge.

Reinitz, J., Mjolsness, E., and Sharp, D.H. (1995). Model for cooperative control of positional information in *Drosophila* by bicoid and maternal hunchback. *J. Exp. Zoo.* 271:47–56.

Savageau, M.A., Voit, E.O., and Irvine, D.H. (1987). Biochemical systems theory and metabolic control theory: 1. Fundamental similarities and differences. *Mathematical Biosciences* 86:127–145.

Simpson, P., Roach, D., Woolley, A., Thorson, T., Johnston, R., Sensabaugh, G., and Mathies, G. (1998). High-throughput genetic analysis using microfabricated 96-sample capillary array electrophoresis microplates. *Proc. Natl. Acad. Sci. USA* 95:2256–2261.

Schwikowski, B., Uetz, P., and Fields, S. (2000). A network of protein-protein interactions in yeast. *Nature Biotech.* 18:1257–1261.

Spellman, P.T., Sherlock, G., Zhang, M.Q., Iyer, V.R., Anders, K., Eisen, M., Brown, P.O., Botstein, D., and Futcher, B. (1998). Comprehensive identification of cell cycle-regulated genes of the yeast *Saccharomyces cerevisiae* by microarray hybridization. *Molecular Biol. Cell.* 9:3273–3297.

Sulston, J.E. and Horvitz, H.R. (1997). Post-embryonic cell lineage of the nematode *Caenorhabditis elegans. Dev. Biol.* 56:110–156.

Sulston, J.E., Schierenberg, E., White, J.G., and Thomson, J.N. (1983). The embryonic cell lineage of the nematode *Caenohabditis elegans. Dev. Biol.* 100:64–119.

Tabara, H., Motohashi, T., and Kohara, Y. (1996). A multi-well version of *in situ* hybridization on whole mount embryos of *Caenorhabditis elegans. Nucleic Acids Research* 24:2119–2124.

Tomita, M., Shimizu, K., Matsuzaki, Y., Miyoshi, F., Saito, K., Tanida, S., Yugi, K., Venter, C., and Hutchison, C. (1999). E-Cell: Software environment for whole cell simulation. *Bioinformatics* 15(1):72–84.

Ueda, H. and Kitano, H. (1998). A generalized reaction-diffusion simulator for pattern formation in biological systems. *Proc. the 6th International Conference on Artificial Life* pp.462–466.

Varma, A. and Palsson, B.O. (1994). Metabolic flux balancing: Basic concepts, scientific and practical use. *Bio/Technology* 12:994–998.

von Bertalanffy, L. (1968). *General System Theory,* Braziler, New York.

von Dassow, G., Meir, E., Munro, E.M., and Odell, G. (2000). The segment polarity network is a robust developmental module. *Nature* 406:188–192.

Watson, J. D. and Crick, F.H. (1953). Molecular structure of nucleic acids: A structure for deoxyribose Nucleic Acid. *Nature* 171:737–738.

White, J.G., Southgate, E., Thomson, J.N., and Brenner, S. (1986). The structure of the nervous system of the nematode *Caenorhabditis elegans. Phil. Trans. R. Soc.* 314:1–340.

Wiener, N., (1948). *Cybernetics or Control and Communication in the Animal and the Machine,* The MIT Press, Cambridge.

Yasuda, T., Bannai, H., Onami, S., Miyano, S., and Kitano, H. (1999). Towards automatic construction of cell-lineage of *C. elegans* from Normarski DIC microscope images. *Genome Informatics* 10:144–154.

Yi, T.-M., Huang, Y., Simon, M., and Doyle, J. (2000). Robust perfect adaptation in bacterial chemotaxis through integral feedback control. *Proc. Natl. Acad. Sci. USA* 97(9):4649–4653.

Part I

Advanced Measurement Systems

2 Automatic Acquisition of Cell Lineage through 4D Microscopy and Analysis of Early *C. elegans* Embryogenesis

Shuichi Onami, Shugo Hamahashi, Masao Nagasaki, Satoru Miyano, and Hiroaki Kitano

Cell lineage analysis is an important technique for studying the development of multicellular organisms. We have developed a system that automatically acquires cell lineages of *C. elegans* from the 1-cell stage up to approximately the 30-cell stage. The system utilizes a set of 4D Nomarski DIC microscope images of *C. elegans* embryo consisting of more than 50 focal plane images at each minute for about 2 hours. The system detects the region of cell nucleus in each of the images, and makes 3D nucleus regions, each of which is a complete set of nucleus regions that represent the same nucleus at the same time point. Each pair of 3D nucleus regions is then connected, if they represent the same nucleus and their time points are consecutive, and the cell lineage is created based on these connections. The resulting cell lineage consists of the three-dimensional positions of nuclei at each time point and their lineage. Encouraged by the performance of our system, we have started systematic cell lineage analysis of *C. elegans*, which will produce a large amount of quantitative data essential for system-level understanding of *C. elegans* embryogenesis.

INTRODUCTION

In the last few decades, biology has been mainly focusing on identifying components that make up the living system. Today, as a result of success in molecular biology and genomics, thousands of genes have been identified, so the focus of biology is now moving toward the next step, understanding how those components work as a whole system. The ultimate goal of this step is the computer simulation of life, that is, reconstruction of living systems on the computer. However, it is still difficult to perform reliable computer simulation even of a single cell.

The first reason for this difficulty is the lack of biological knowledge. Based on the genomic sequence, the number of human genes was predicted as approximately 26,000 (Venter et al., 2001), and about 4300 genes were predicted even for *Eschericia coli*, a single-cellular prokaryote (Barrick et al., 1994). However, with some exceptions, information on the functions of those genes is quite limited. In order to determine those functions efficiently, automation of biological experiments is necessary. Such automation, following on from the automatic DNA sequencer, the DNA microarray system, etc., will greatly increase the quality of computer simulation.

The second reason is the lack of quantitative information. Historically, biology has been mainly accumulating qualitative information, such as "the expression of gene increases" and "the nucleus moves to the anterior." However, for computer simulation, quantitative information is necessary, such as "the expression of gene increases at v ng/s," and "the position of the nucleus is (x, y, z)." To obtain such quantitative information, precisely controlled analytical instruments need to be developed.

The third reason is the immaturity of modeling technology and simulation technology. Several software packages have been developed for biological computer simulation (Mendes, 1993; Morton-Firth and Bray, 1998; Tomita et al., 1999; Shaff and Loew, 1999; Kyoda et al., 2000). These efforts have greatly improved modeling and simulation technology for simple biological processes, such as reactions among free molecules, and so the accuracy and reliability of computer simulation have been greatly increased for single-cellular organisms and individual cells. However, almost no technology has been developed for more complicated biological processes, such as sub-cellular localization of molecules and organelles, cell division, and three-dimensional positioning of cells. These technologies are essential for reliable simulation of multicellular organisms.

This chapter reviews our automatic cell lineage acquisition system, which is one of our approaches we have developed for computer simulation of *C. elegans*. The system automates biological experiments and produces quantitative data. The end of this chapter briefly reviews our other approaches, which are improving modeling and simulation technology, and then briefly overviews our approaches as a whole.

THE NEMATODE, *C. ELEGANS*

There are good introductions to *C. elegans* in other literatures (Wood et al., 1988; Riddle et al., 1997), so a detailed introduction of this organism is omitted. Briefly, *C. elegans* is the simplest multicellular organism that has been most extensively analyzed in molecular and developmental biology. This organism is also the first multicellular organism whose genome sequence has been completely identified (The C. elegans Sequencing Consortium, 1998), and is leading the other multicellular experimental organisms in post genome sequencing analysis, such as functional genomics and proteomics. Thus, *C. elegans* is expected to be the first multicellular organism whose life is fully reconstructed on the computer.

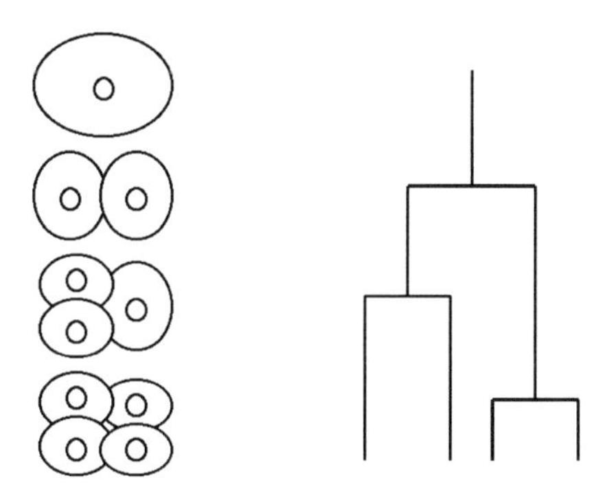

Figure 2.1 Cell lineage. When the fertilized egg undergoes a series of cell divisions shown on the left, the cell lineage is described as shown on the right. In the cell lineage, the vertical axis represents the time and the horizontal axis represents the direction of division (left-right and anterior-posterior).

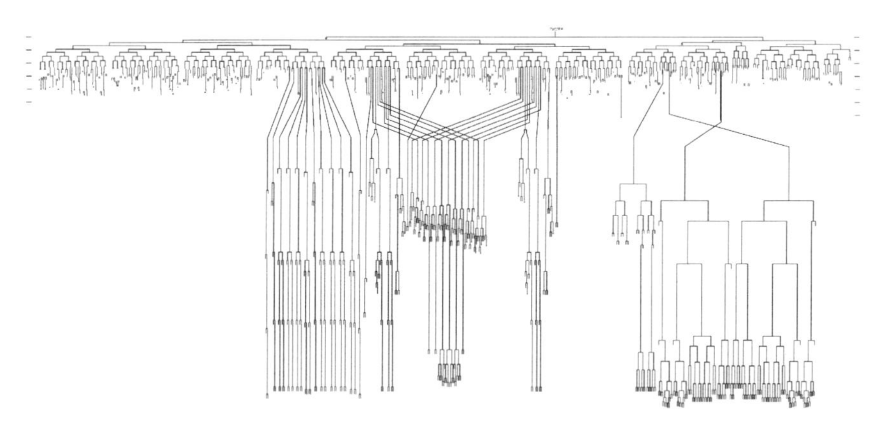

Figure 2.2 The complete cell lineage of *C. elegans* (Sulston et al., 1983).

CELL LINEAGE AND ITS APPLICATION

Generally, a multicellular organism is a mass of cells that are generated from a single cell – i.e. the fertilized egg – through successive cell divisions. Each cell division is a process whereby a single mother cell produces a pair of daughter cells. Cell lineage is a tree-like description of such mother-daughter relationships starting from the fertilized egg (in a wide sense, starting from a specific cell) (Moody, 1999) (Figure 2.1). It usually includes information on the timing and the direction of each cell division. The complete cell lineage – from the fertilized egg to the adult – has been identified for several simple multicellular organisms, such as *C. elegans* (Sulston et al., 1983) and *Halocynthia roretzi* (Nishida, 1987)(Figure 2.2).

The most typical application of cell lineage is gene function analysis

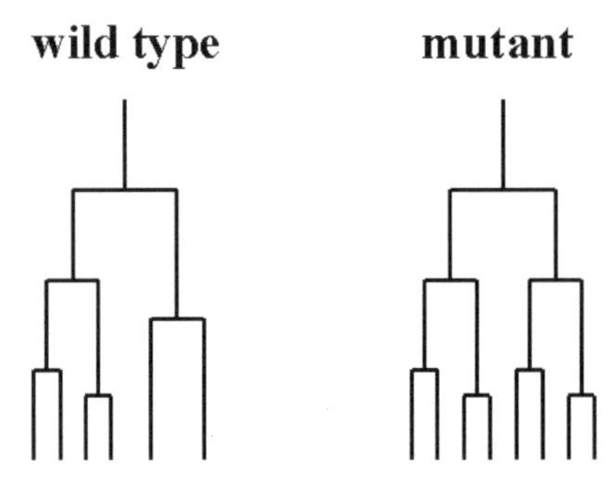

Figure 2.3 Comparison of cell lineage between wild type and mutant animals. When the wild type and the mutant cell lineages are described as in this figure, the mutated gene plays some roles in the differentiation of the two daughter cells produced at the first cell division

by comparing cell lineages among wild type and mutant animals (Figure 2.3). Through such analysis, many gene functions have been uncovered. So, cell lineage analysis is an important technique for studying the development of multicellular organisms, as well as *in situ* hybridization, immunohistochemistry, and GFP-fusion gene expression.

HISTORY OF CELL LINEAGE ANALYSIS PROCEDURE

This section reviews the history of the cell lineage analysis procedure, focusing on the procedure for *C. elegans*. But with some difference in the dates, the history is almost the same in other animals.

The entire cell lineage of *C. elegans* was reported by Sulston *et al.* in 1983 (Sulston et al., 1983). In this work, they used a rather primitive procedure whereby they directly observed the animal through a Nomarski DIC microscope and sketched it. A Nomarski DIC microscope visualizes subtle differences of thickness and refraction index in the light path, and has an advantage that the intra-cellular structure of living transparent cells can be studied without staining (Spector et al., 1998). Through this microscope, moving the focal plane up and down, Sulston *et al.* observed and sketched a 14-hour process of *C. elegans* embryogenesis, from fertilization to hatching (Figure 2.4), which must have been quite laborious.

The four-dimensional microscope imaging system (4D microscope), developed by Hird *et al.* in 1993, greatly reduced the laboriousness of cell lineage analysis (Hird and White, 1993). By controlling the focusing device and the camera, the system automatically captures microscope images of different focal planes, and repeats the process with a given interval. This system obtains a set of microscope images that contain the 3D structure information of an embryo at different time points starting

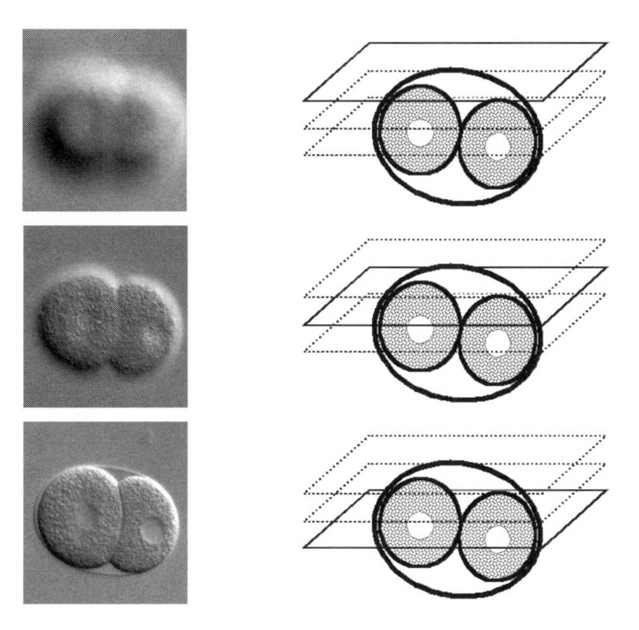

Figure 2.4 Nomarski DIC microscope images of different focal planes. Nomarski DIC microscope images of a 2-cell stage embryo are shown. Moving the focal plane up and down, the 3D structure of the embryo can be recognized.

from fertilization (Figure 2.5). Then, those images are closely analyzed to derive the cell lineage. A GUI supporting tool, developed by Schnabel *et al.* in 1997, further reduced the laboriousness of cell lineage analysis (Schnabel et al., 1997).

As is reviewed above, cell lineage analysis has become much easier than that in Sulston's era, but it is still quite laborious. The number of mutants whose cell lineage is identified, is quite small compared with the number of mutants whose responsible gene is identified and sequenced.

AUTOMATIC CELL LINEAGE ACQUISITION

We are developing a system that automatically acquires cell lineages of *C. elegans* (Yasuda et al., 1999). The latest version of our system has the ability to acquire the cell lineage from the 1-cell stage up to approximately the 30-cell stage (Hamahashi et al., unpublished). This section reviews the process of our system.

The system utilizes a set of 4D Nomarski DIC microscope images to extract the cell lineage (Figure 2.5). The 4D microscope system is able to capture more than 50 images per minute, changing the focal plane position by 0.5 μm for each image. With this system, a set of multi-focal plane images of a *C. elegans* embryo is captured every minute for about

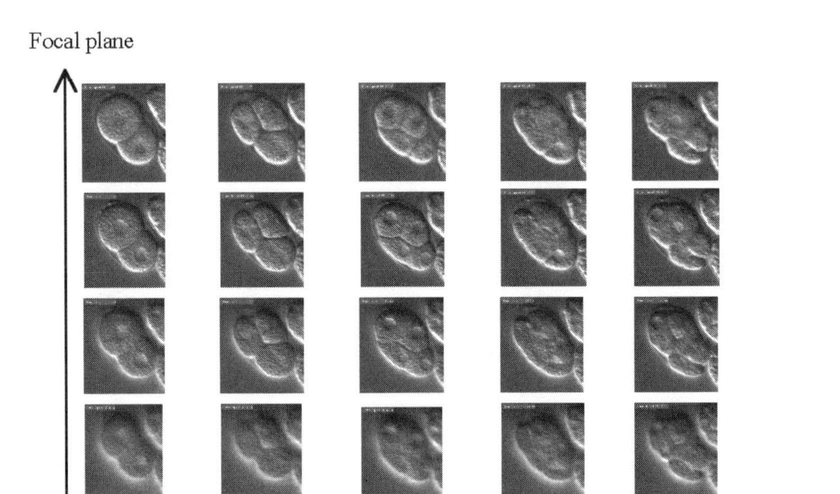

Figure 2.5 4D microscope images.

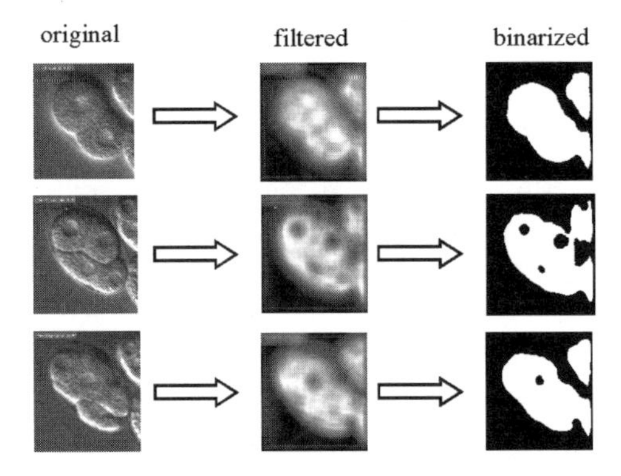

Figure 2.6 Example of an nucleus detection filter.

2 hours. Since the height of the embryo is about 25 μm, each multi-focal plane image includes all 3D structure information of the embryo at the corresponding time point.

The system then processes each of the images captured in the previous step, and detects the regions of cell nucleus in the image (Figure 2.6). In the Nomarski microscope image, the region of cytoplasm looks bumpy as a result of the existing organelles, such as lysosome and mitochondria. On the other hand, the nucleus region, without those organelles, looks smooth. We found that several basic image-processing filters (i.e. Kirsch's edge detection filter (Kirsch, 1971), entropy filter (Jahne et al., 1999), etc.) efficiently detect those nucleus regions (Yasuda et al., 1999; Hamahashi et

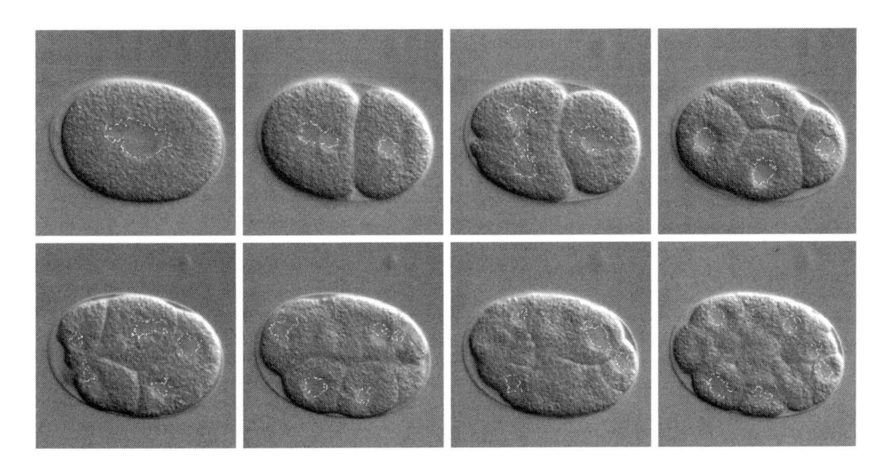

Figure 2.7 Detected nucleus regions. Each of the detected nucleus regions is enclosed by a white line.

al., unpublished). Several new filters applicable to this nucleus detection were also developed (Yasuda et al., 1999; Hamahashi et al., unpublished). By appropriately combining those filters, we established an excellent nucleus detection algorithm (Hamahashi et al., unpublished). With this algorithm, non-error nucleus detection is carried out from the 1-cell stage to about the 30-cell stage (Figure 2.7).

In wild type embryo, the diameter of a nucleus is about 7 μm at the 2-cell stage and 4.5 μm at the 20-cell stage, whereas our system captures microscope images every 0.5 μm of focal plane position. Therefore, at every time point, a nucleus is detected on several different focal planes. In the next step, the system makes 3D nucleus regions, each of which is a complete set of nucleus regions that represent the same nucleus at the same time point. Then, the system connects each pair of 3D nucleus regions, if they represent the same nucleus and their time points are consecutive. In the previous two steps, a pair of nucleus regions is recognized as representing the same nucleus, when one nucleus region overlaps the other either on the same focal plane at the next time point or on the next focal plane at the same time point. As the result, the lineage of 3D nucleus regions is recognized through out the entire period of the 4D microscope images.

Finally, the cell lineage is created based on the above 3D nucleus region lineage. The system calculates the centroid position of each 3D nucleus region, and outputs those centroid positions and their lineage (Figure 2.8).

This section briefly reviews the process of our cell lineage detection system. The current system utilizes our Beowulf PC cluster (Okuno et al., 2000), made up of 32 PCs, to execute all the above processes except 4D microscope image recording, and within 9 hours, can deduce the cell lineage up to the 30- to 40-cell stage after setting the 4D microscope images

```
true-010-043-0005 (29.82, 30.555, 21.5)
true-010-043-0005 -> 010-038-0011 (29.715, 30.24, 21.5)
010-038-0011 -> 011-038-0023 (30.66, 29.085, 21.5)
011-038-0023 -> 012-038-0011 (31.5, 27.51, 21)
012-038-0011 -> 013-037-0019 (34.02, 25.2, 21)
013-037-0019 -> 014-038-0010 (34.02, 24.885, 21)
014-038-0010 -> 015-038-0005 (39.375, 17.22, 20.5)
014-038-0010 -> 015-038-0022 (29.505, 31.815, 21)
015-038-0022 -> 016-037-0023 (31.92, 28.035, 21)
016-037-0023 -> 017-037-0010 (32.865, 27.405, 21)
017-037-0010 -> 018-037-0016 (26.04, 37.065, 20.5)
017-037-0010 -> 018-038-0013 (38.85, 20.055, 20.5)
017-037-0010 -> 018-041-0003 (32.34, 31.5, 22)
018-037-0016 -> 019-037-0014 (25.095, 36.33, 21)
018-038-0013 -> 019-036-0003 (39.06, 22.575, 22)
018-041-0003 -> 019-042-0011 (32.865, 32.13, 23)
019-037-0014 -> 020-037-0025 (24.675, 35.91, 20.5)
019-036-0003 -> 020-036-0011 (41.895, 17.535, 18.5)
019-036-0003 -> 020-040-0006 (35.28, 27.825, 23)
019-036-0003 -> 020-042-0007 (39.69, 16.485, 21)
019-036-0003 -> 020-044-0009 (39.165, 15.96, 22)
020-036-0011 -> 021-037-0008 (32.97, 29.505, 22.5)
021-037-0008 -> 022-038-0007 (33.705, 28.665, 22.5)
022-038-0007 -> 023-038-0006 (34.335, 28.035, 22.5)
023-038-0006 -> 024-036-0011 (34.86, 27.615, 22.5)
024-036-0011 -> 025-035-0005 (35.28, 27.615, 22)
025-035-0005 -> 026-035-0003 (35.49, 27.825, 22)
026-035-0003 -> 027-035-0007 (35.595, 27.93, 22)
027-035-0007 -> 028-035-0011 (39.27, 21.525, 21.5)
027-035-0007 -> 028-039-0011 (32.76, 33.285, 23)
028-035-0011 -> 029-034-0007 (39.48, 21.84, 21.5)
028-039-0011 -> 029-039-0007 (31.815, 34.125, 22.5)
029-034-0007 -> 030-034-0013 (39.69, 22.47, 21.5)
029-039-0007 -> 030-036-0017 (29.715, 34.335, 22)
030-034-0013 -> 031-034-0008 (39.69, 22.155, 21.5)
030-036-0017 -> 031-033-0009 (29.61, 34.65, 22)
031-034-0008 -> 032-034-0010 (39.9, 21.63, 21.5)
031-033-0009 -> 032-033-0010 (29.505, 33.915, 22)
032-034-0010 -> 033-034-0008 (40.11, 20.475, 21.5)
032-033-0010 -> 033-033-0014 (29.61, 34.02, 22)
033-034-0008 -> 034-033-0011 (40.215, 19.53, 21.5)
033-033-0014 -> 034-033-0016 (29.82, 32.76, 22)
034-033-0011 -> 035-033-0008 (34.545, 25.41, 21.5)
035-033-0008 -> 036-031-0019 (27.3, 36.015, 20)
035-033-0008 -> 036-033-0009 (40.005, 18.165, 20.5)
```

Figure 2.8 Text data for *C. elegans* cell lineage.

(Hamahashi et al., unpublished). We also developed a software package that three-dimensionally visualizes the resulting lineage data (Hamahashi et al., unpublished), which may help three-dimensional understanding of nucleus movement and division (Figure 2.9). Moreover, with this package, lineages of two different individuals – e.g., wild type and mutant – can be visualized on the same screen.

ADVANTAGES OF AUTOMATIC CELL LINEAGE ACQUISITION SYSTEM

As noted in the previous section, we have successfully developed a high-performance automatic cell lineage acquisition system. The advantages of this system are outlined below.

The most significant advantage of this system is automation, as can easily be imagined from the contribution of the automatic DNA sequencer to biology. The required human effort for our system is almost the same as that of the DNA sequencer. The processing time of 9 hours is almost the same as that of the sequencer in its early days. With this system, large scale and systematic cell lineage analysis is made possible.

The second advantage is quantitative data production. The three-dimensional nucleus position at each time point which the system outputs is quantitative data, which is essenaital for simulation studies, especially when simulation models are developed and simulation results are analyzed. Our system can be applied to many individual animals, wild types and mutants, and the resulting data will greatly improve the accuracy of computer simulation.

The third advantage is the reproducibility of the results. When cell lin-

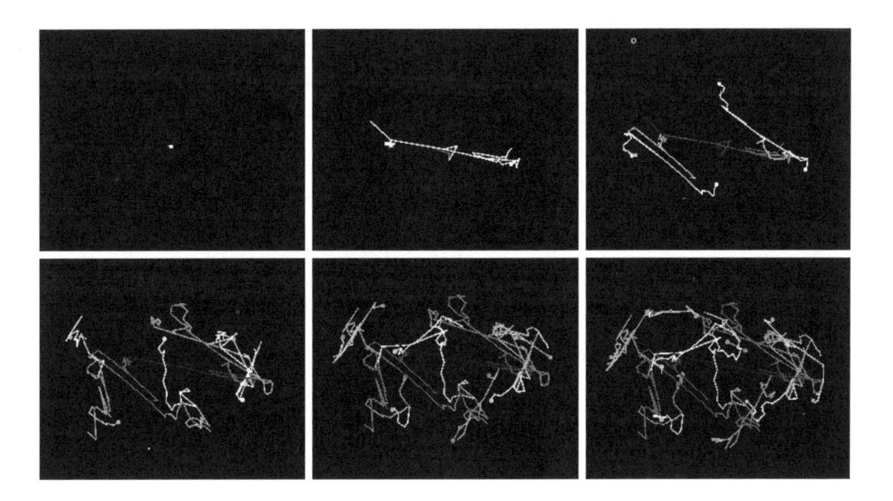

Figure 2.9 Three-dimensional view of a *C. elegans* cell lineage. The centroid positions of 3D nucleus regions are traced from 1-cell to 19-cell stage. The white circles represent the centroid positions at the viewing time point. On this viewer, it is possible to freely change the viewing time point forward and backward, and also rotate the viewing point three-dimensionally.

eages are manually analyzed, the resulting cell lineage is unreproducible. For example, the definitions of nucleus position and cell division time may vary depending on who made the analysis and, even if the same people made it, when it was done. In our system, such definitions are exactly the same through all individual measurements and the results are completely reproducible. The results are thus suitable for statistical analysis, such as calculating the mean, variance, standard error, etc.

Fourthly, the system is applicable to other organisms. The basis of the system is an image-processing algorithm for Nomarski microscope images. Thus, in principle, the system is applicable to all transparent cells and embryos. Future application to other organisms, such as *Halocynthia roretzi*, mouse, is promising.

Finally, the system offers complementarity of cell lineage data. As a result of success in molecular biology and genomics, a variety of large-scale analyses are currently undertaken, such as DNA mircoarray, protein chip, and systematic *in situ* hybridization. However, combinations of those analyses are not so fruitful since they all measure the same object – gene expression level. Cell lineage data is quite complementary to gene expression data, therefore, the combination of our cell lineage analysis with gene expression analyses will provide useful information for biology.

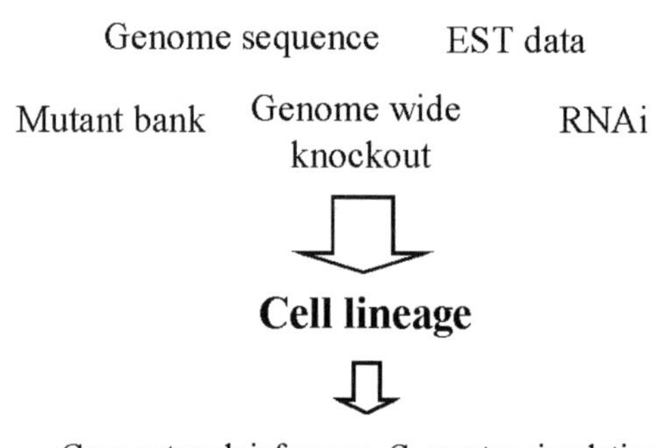

Figure 2.10 Systematic cell lineage analysis of *C. elegans*.

FUTURE DEVELOPMENT OF THE CELL LINEAGE ACQUISITION SYSTEM

As described in this chapter, the current version of our system extracts a *C. elegans* cell lineage of up to the 30-cell stage in 9 hours.

The biggest challenge in the future system development is, of course, to extend the applicable embryonic period, up to the 100-cell stage, 200-cell stage, and beyond. The current limit of the applicable period is imposed by the performances of the 4D microscope system, such as the speed of the z-axis driving motor and the image-capturing period of the CCD camera. The performance of these devices is rapidly being improved, so such device-dependent limit will likely be overcome in the near future. The limit of the current algorithm may be around the 60-cell stage. For the later stages, an improved algorithm will be required. Nucleus detection is quite difficult even for humans after the 100-cell stage, so for later stages, GFP-labeling of nucleus or other nucleus labeling techniques may be necessary.

Shortening the processing time is another important challenge, but the solution seems relatively easy. Dramatic improvement of CPU speed will greatly shorten the processing time of our system.

SYSTEMATIC CELL LINEAGE ANALYSIS

Encouraged by the performance of the current cell lineage acquisition system, we have started systematic cell lineage analysis of *C. elegans* embryo (Figure 2.10).

As the first step, we are currently accumulating many wild type cell lineages in order to establish the standard wild type cell lineage, which describes the mean value of nucleus position at each time point together with some statistical data, such as the variance, error distribution, etc. As well as wild type animals, we are also analyzing cell lineages for many mutants that are already known to play important roles in early embryogenesis. By analyzing the results, we will confirm and also improve the current understanding of early embryogenesis.

In *C. elegans*, there is a quite well organized mutant-stocking system (Caenorhabditis Genetic Center [1]). In addition, several whole genome knock-outing projects are being undertaken either by efficient mutagenesis (Gengyo-Ando and Mitani, 2000) or *RNAi* (Fraser et al., 2000), taking advantage of the complete genome sequence data (The C. elegans Sequencing Consortium, 1998). We are planning to start a systematic cell lineage analysis for those knock-out animals in future. The resulting data, together with the systematic gene expression data (Tabara et al., 1996), will provide useful information for the complete understanding of *C. elegans*.

COMPUTER SIMULATION OF *C. ELEGANS*

Our cell lineage system can produce a large amount of quantitative data, which is useful for computer simulation. To achieve computer simulation of *C. elegans* embryogenesis, the authors are also running several other closely related projects, as outlined below.

The quality of computer simulation is largely dependent on the simulation model, thus the model construction is an important process in simulation studies. To help this process, we are developing gene regulatory network inference methods. An efficient method has been developed for large-scale gene expression data, such as DNA microarray data (Kyoda et al., 2000). Currently, we are developing a sophisticated gene network modeling scheme based on this method, and are also trying to develop a gene network inference method that utilizes the cell lineage information.

For improving the technology of biological computer simulation, we are developing a model description language for biological computer simulation (Nagasaki et al., 1999), named *bio-calculus*. The language will be able to describe a variety of biological processes observed in multicellular organisms, such as sub-cellular positioning of molecules and organelles, cell division, and three-dimensional positioning of cells. We are also developing several software packages so that a variety of biological models described using the language can be executed (Nagasaki et al., 1999). Currently, a very early period of *C. elegans* embryo is being modeled and simulated (Nagasaki et al., unpublished) in order to improve the applicability of the language and its software packages (Figure 2.11, 2.12). In future, utilizing our cell lineage information, we will gradually refine our *C. elegans* model and extend the target period, to improve the modeling and simulation technology further.

---

1 `http://biosci.umn.edu/CGC/CGChomepage.htm`

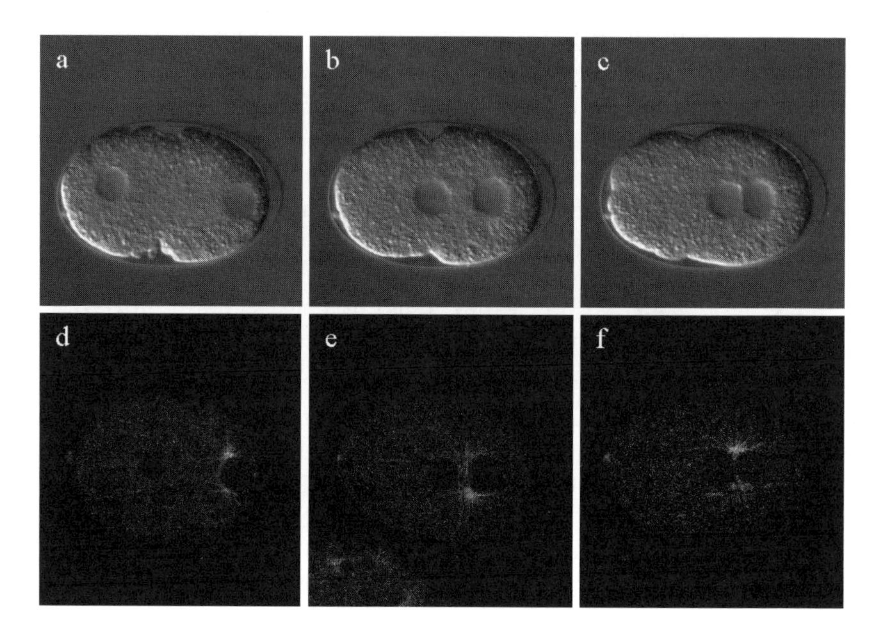

Figure 2.11 Pronucleus movement of *C. elegans* embryo. a)–c) Nomarski DIC microscope images of very early *C. elegans* embryo just after fertilization. The anterior is left. The oocyte pronucleus (left) and the sperm pronucleus (right) move toward each other and finally meet in the posterior hemisphere. The movement of the sperm pronucleus mainly depends on microtubules (MTs) (Hird and White, 1993). d)–f) Confocal microscope images of *C. elegans* embryo stained with MT specific antibody. MTs are growing from the centrosomes on the sperm pronucleus.

CONCLUSION

This chapter reviewed our automatic cell lineage acquisition system. The system will produce a large amount of quantitative data, which is valuable for computer simulation, though the data are still insufficient for the complete *C. elegans* simulation. We must therefore keep developing new experimental technologies. It is hoped that all our approaches will functionally work together to enable us to achive the ultimate goal – the computer simulation of life.

ACKNOWLEDGEMENT

The authors thank Nick Rhind for his cell lineage drawing.

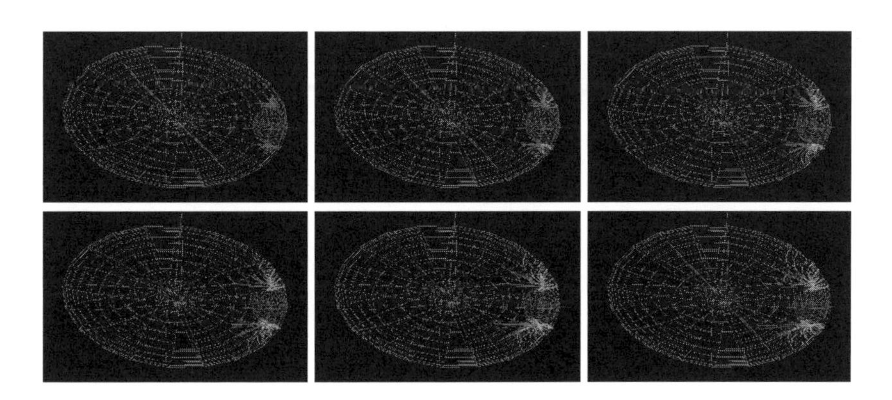

Figure 2.12 Computer simulation of MT dependent sperm pronucleus movement in *C. elegans* embryo. The small circle represents the sperm pronucleus and the white lines growing from the sperm pronucleus represents MTs growing from the centrosomes on the pronucleus.

References

Barrick, D., Villanueba, K., Childs, J., Kalil, R., Schneider, T.D., Lawrence, C.E., Gold, L., and Stormo, G.D. (1994). Quantitative analysis of ribosome binding sites in *E. coli*. *Nucleic Acids Res.* 22:1287–1295.

The C. elegans Sequencing Consortium. (1998). Genome sequence of the nematode C. elegans. a platform for investigating biology. *Science* 282:2012–2018.

Fraser, A., Kamath, R.S., Zipperlen, P., Martinez-Campos, M., Sohrmann, M., and Ahringer, J. (2000). Functional genomic analysis of *C. elegans* chromosome I by systematic RNA interference. *Nature* 408:325–330.

Gengyo-Ando, K. and Mitani, S. (2000). Characterization of mutations induced by ethyl methanesulfonate, uv, and trymethylpsoralen in the nematode *Caenorhabditis elegans*. *Biochem. Biophys. Res. Comm.* 269:64–69.

Hamahashi, S., Onami, S., and Kitano, H. Unpublished.

Hird, S. and White, J.G. (1993). Cortical and cytoplasmic flow polarity in early embryonic cells of *Caenorhabditis elegans*. *J. Cell Biol.* 121:1343–1355.

Jahne, B., Haussecker, H., and Geissler, P. editors. (1999). *Handbook of computer vision and applications*. Academic Press.

Kirsch, R. (1971). Computer determination of the constituent structure of biological images. *Comput. Biomed. Res.* 4:315–328.

Kyoda, K.M., Morohashi, M., Onami, S., and Kitano, H. (2000). A gene network inference method from continuous-value gene expression data of wild-type and mutants. *Genome Informatics*, 11:196–204.

Kyoda, K.M., Muraki, M., and Kitano, H. (2000). Construction of a generalized simulator for multi-cellular organisms and its application to smad signal transduction. *Proc. Pacific Symposium on Biocomputing 2000* pp.317–328.

Mendes, P. (1993). Gepasi: A software package for modelling the dynamics, steady states and control of biochemical and other systems. *Comput. Applc. Biosci.* 9:563–571.

Moody, S.A., editor. (1999). *Cell lineage and fate determination*. Academic Press.

Morton-Firth, C.J., and Bray, D. (1998). Predicting temporal fluctuations in an intracellular signalling pathway. *J. Theor. Biol.* 192:117–128.

Nagasaki, M., Miyano, S., Onami, S., and Kitano, H. Unpublished.

Nagasaki, M., Onami, S., Miyano, S., and Kitano, H. (1999). Bio-calculus: Its concept and molecular interaction. *Genome Informatics* 10:133–143.

Nishida, H. (1987). Cell lineage analysis in ascidian embryos by intracellular injection of a tracer enzyme. III. up to the tissue restricted stage. *Dev. Biol.* 121:526–541.

Okuno, H.G., Kyoda, K.M., Morohashi, M., and Kitano, H. (2000). Initial assessment of ERATO-1 beowulf-class cluster. *Proc. Parallel and distributed computing for symbolic and irregular applications* pp.372–383.

Riddle, D.L., Blumenthal, T., Meyer, B.J., and Priess, J.R., editors. (1997). *C. elegans II*. Cold Spring Harbor Laboratory Press.

Schaff, J. and Loew, L.M. (1999). The virtual cell. *Proc. Pacific Symposium on Biocomputing '99* pp.228–239.

Schnabel, R., Hutter, H., Moerman, D., and Schnabel, H. (1997). Assessing normal embryogenesis in *Caenorhabditis elegans* using 4D microscope: variability of development and regional specification. *Dev. Biol.* 184:234–265.

Spector, D.L., Goldman, R.D., and Leinwand, L.A. editors. (1988). *Cells - A Laboratory Manual*. Cold Spring Harbor Laboratory Press.

Sulston, J.E., Shierenberg, E., White, J.G., and Thomson, J.N. (1983). The embryonic cell lineage of the nematode *Caenorhabditis elegans*. *Dev. Biol.* 100:64–119.

Tabara, H., Motohashi, T., and Kohara, Y. (1996). A multi-well version of *in situ* hybridization on whole mount embryos of *Caenorhabditis elegans*. *Nucleic Acids Res.* 24:2119–2124.

Tomita, M., Hashimoto, K., Takahashi, K., Shimizu, T.S., Matsuzaki, Y., Miyoshi, F., Saito, F., Tanita, S., Yugi, K., Vender, J.C., and Hutchison, C.A. (1999). E-cell: software environment for whole-cell simulation. *Bioinformatics* 15:72–84.

Venter, J.C., et al. (2001). The sequence of the human genome. *Science* 291:1304–1351.

Wood, W.B. and the Community of C. elegans Researchers, editors. (1988). *The nematode Caenorhabditis elegans*. Cold Spring Harbor Laboratory.

Yasuda, T., Bannai, H., Onami, S., Miyano, S., and Kitano, H. (1999). Towards automatic construction of cell lineage of *C. elegans* from Nomarski DIC microscope images. *Genome Informatics* 10:144–154.

Part II

Reverse Engineering and Data Mining from Gene Expression Data

3 The DBRF Method for Inferring a Gene Network from Large-Scale Steady-State Gene Expression Data

Shuichi Onami, Koji M. Kyoda, Mineo Morohashi, and Hiroaki Kitano

Complete genome sequence has enabled whole-genome expression profiling and genome deletion projects, which are generating large-scale gene expression profiles corresponding to hundreds of deletion mutants. To obtain valuable information from those profiles is an important challenge in current biology. This chapter reviews the Difference-Based Regulation Finding (DBRF) method, which infers the underlying gene network from those profiles. The method 1) infers direct and indirect gene regulations by interpreting the difference of gene expression level between wild-type and mutant, and 2) eliminates the indirect regulations. One of the major characteristics of the method is its applicability to continuous-value expression data, whereas the other existing method can only deal with binary data. The performance of the method was evaluated using artificial gene networks by varying the network size, indegree of each gene, and the data characteristics (continuous-value or binary). The results showed that the method is superior to the other methods. The chapter also reviews the applicability of the DBRF method to real gene expression data. The method was applied to a set of yeast DNA microarray data which consisted of gene expression levels of 249 genes in each of single gene deletion mutants for the 249 genes. In total, 628 gene regulatory relationships were inferred, where the accuracy of the method was confirmed in MAP kinase cascade. The DBRF method will be a powerful tool for genome-wide gene network analysis.

INTRODUCTION

Recent progress in the field of molecular biology enables us to obtain huge amounts of data. The rapidly increasing amount of known sequence data, or massive gene expression data, requires computational effort to extract information from them. So far, much attention has been focused on developing various advanced computational tools, such as for homology search, protein classification, gene clustering, and so forth.

Several significant studies have attempted to establish a method to infer a gene regulatory network from large-scale gene expression data. The gene expression data are primarily obtained as either 1) time series, or 2) steady-state data. For analyzing the time series, networks are inferred by employing various techniques (e.g., information theory (Liang et al., 1998), genetic algorithms (Morohashi and Kitano, 1999), or simulated annealing (Mjolsness et al., 1999)). One of the shortcomings of the time series approach is that it requires experimental data that are taken at very short intervals and are almost free from experimental noise. These requirements are almost impossible to meet with current techniques.

On the other hand, some methods have already been proposed for inferring regulatory relationships using steady-state gene expression data. The steady-state data can be obtained by altering specific gene activities, such as knock-outing or overexpressing genes. Gene knock-outing is currently being developed on a large scale for a variety of experimental animals, such as *S. cerevisiae* (Winzeler et al., 1999; Hughes et al., 2000), *C. elegans* (Gengyo-Ando and Mitano, 2000), and *Drosophila* (Spradling et al., 1999), by which various gene expression profiles will be produced in a unified manner. Moreover, the discovery of RNA interference enables us to create gene knockout animals easily and is applicable to *C. elegans* and *Drosophila* (Sharp, 1999). Akutsu et al. (1998) calculated upper and lower bounds on the number of experiments that would be required if the network were Boolean. More recently, Ideker et al. (2000) proposed an inference method called *predictor*. The predictor method provides candidate networks represented by a Boolean network that are consistent with expression data by employing combinatorial optimization techniques.

A drawback of these methods is that they assume a gene network as a Boolean network where the expression levels are represented as binary values. In general, experimental data have continuous values, and thus the data should be translated into binary data in order to apply the methods. Such translation may cause the data to lack the information needed to infer regulatory relationships. If binary data are used, even 3-state (e.g., wild-type, deletion, and overexpression) levels may be impossible to be represented, in which case the underlying inherent regulatory relationships cannot be accurately represented.

In this chapter, we review the Difference-Based Regulation Finding (DBRF) method which is a gene network inference method using steady-state gene expression data (Kyoda et al., 2000). The DBRF method is applicable to expression data represented as not only binary values, but also continuous values. The chapter is organized as follows: in the next section, we review the algorithm of the DBRF method. In the third section, the performance of the DBRF method is reviewed. The performance was studied using artificial gene regulatory networks. In the fourth section, we review our application of this method to yeast DNA microarray data. In the fifth section, we discuss the advantages and characteristics of this method.

THE DIFFERENCE-BASED REGULATION FINDING METHOD

We describe the DBRF method for inferring a gene regulatory network from the steady-state gene expression data of wild-type and deletion/overexpression mutants (Kyoda et al., 2000). An example of interaction matrix I is shown in Figure 3.1(a), which represents gene interactions. Rows of I represent the genes that regulate the genes in columns (e.g., a_0 activates both a_2 and a_3, and a_2 represses a_3). We assume that the data are given by an expression matrix E, a set of observed steady-state gene expression levels for all genes over all mutation experiments. An example of E is shown in Figure 3.1(b). Rows of E represent the deleted genes while columns represent the steady-state expression levels in each gene. We apply the method to the expression matrix E in order to derive the interaction matrix I.

The basic procedure of the DBRF method involves two steps: 1) infer direct and indirect regulations among the genes from expression data, and 2) eliminate the indirect regulations from the above regulations to infer a parsimonious network.

	a_0	a_1	a_2	a_3
a_0			+	+
a_1			+	
a_2				−
a_3				

(a) The interaction matrix I

	x_0	x_1	x_2	x_3
wt	3.750	3.750	8.939	0.078
a_0^-	–	3.750	8.769	0.011
a_1^-	3.750	–	8.769	0.086
a_2^-	3.750	3.750	–	5.476
a_3^-	3.750	3.750	8.939	–

(b) The expression matrix E

Figure 3.1 An example of the matrices which show the regulations and steady-state data of a network. (a) An interaction matrix I. (b) An expression matrix E. The values in the matrix are calculated by model equations shown in Figure 3.3(b).

Inference of a Redundant Gene Regulatory Network

A simple way to determine the regulatory relationships between genes is to see the difference of expression level [1] between wild-type (wt) and mutant data. In the first step, the DBRF method derives the relationships between genes as such. The gene regulatory relationship is inferred according to the rule shown in Table 3.1. It is clear that gene a activates (represses) the expression of gene b if the expression level of gene b goes

1 'expression level' is represented as absolute or relative quantities of mRNA or proteins.

down (up) when gene a is deleted. The computational cost of this comparison process is $O(n^2)$.

This process can infer not only direct gene regulations but also indirect ones. For example, the process infers a gene interaction from gene a_1 to gene a_3, since the expression level x_3 is different between wt and a_1^- (Figure 3.1(b)). However, this interaction is an indirect gene interaction through gene a_2 (Figure 3.3(a)). In the subsequent process, these indirect gene regulations are eliminated, and a parsimonious gene regulatory network is inferred.

Table 3.1 Inference rule of genetic interaction between gene a and gene b from steady-state gene expression data of wild-type, single deletion and overexpression mutant.

		Expression level of gene b	
		up	*down*
Gene a	deletion	$a \dashv b$	$a \to b$
	overexpression	$a \to b$	$a \dashv b$

Inference of a Parsimonious Gene Regulatory Network

This step infers a parsimonious gene regulatory network from the redundant gene regulatory network inferred above by eliminating indirect edges (gene regulations). In order to eliminate those indirect edges, for each pair of genes, we 1) find out whether there are more than one route between those genes, 2) check whether regulatory effects (activation/inactivation) of those routes are the same, and 3) eliminate redundant routes if the effects are the same.

Figure 3.2 shows the algorithm for inferring a parsimonious gene regulatory network. In order to develop 1) and 2), we modified Warshall's algorithm (Gross and Yellen, 1999). Warshall's algorithm is based on the transitive rule that there is an edge from a_i to a_k if edges from a_i to a_j, and from a_j to a_k exist. For example, if there are edges from a_i to a_k, from a_i to a_j, and from a_j to a_k, the algorithm finds out that there are two routes from a_i to a_k. Even if a route consists of more than three genes, 1) can be done using this algorithm. 2) is implemented by adding a function counting the number of negative regulations in each route to Warshall's algorithm. The regulatory effect only depends on the parity of the number of negative regulations involved in the route (Thieffry and Thomas, 1998). For example, given two routes connecting the same pair of genes, the regulatory effects of those two routes are the same if the parities of that number are the same. The number of negative regulations is counted in each route found in 1), and groups of routes whose regulatory effects are the same, are detected in the algorithm. For 3), we define that if there

procedure

var *input*: an n-node gene regulatory network G with node $a_1, a_2, \ldots, a_n$.
output: the transitive closure of gene regulatory network G.
tn: total number of negative regulations.

begin

initialize gene regulatory network G_0 to be network G.
for $i = 1$ to n **do**
  for $j = 1$ to n **do**
    if (a_j, a_i) is an edge in network G_{i-1}
      for $k = 1$ to n **do**
        if (a_i, a_k) is an edge in network G_{i-1}
          $tn = (a_j, a_i)_{negative\_num.} + (a_i, a_k)_{negative\_num.}$
          if $(a_j, a_k)_{negative\_num.}$ is even, and tn is even.
            eliminate edge (a_j, a_k) to G_{i-1}.
            $(a_j, a_k)_{negative\_num.} = tn$.
          if $(a_j, a_k)_{negative\_num.}$ is odd, and tn is odd.
            eliminate edge (a_j, a_k) to G_{i-1}.
            $(a_j, a_k)_{negative\_num.} = tn$.
return gene regulatory network G_n

end

Figure 3.2 An algorithm for inferring a parsimonious gene regulatory network. Here let G be an n-node digraph with nodes $a_1, a_2, \ldots, a_n$. This algorithm constructs a sequence of digraphs, $G_0, G_1, \ldots, G_n$, such that $G_0 = G$, G_i is a subgraph of G_{i-1}, $i = 1, \ldots, n$ because of eliminating redundant edges subsequently. (a_p, a_q) is the p-q element of the interaction matrix I. Each element of the interaction matrix I has the storage for total negative regulation number between gene p and gene q.

is more than one possible route between a given pair of genes and their regulatory effects are the same, the route consisting of the largest number of genes is the parsimonious route and the others are redundant. Thus, for each pair of genes, the number of genes in each route of the same effect is counted, and all but the one consisting of the largest number of genes are eliminated in the algorithm. The computational cost of this algorithm is $O(n^3)$.

COMPUTATIONAL EXPERIMENTS

Since the experimental data of deletion mutants are being produced by several yeast genome deletion projects (Winzeler et al., 1999; Hughes et al., 2000), it is reasonable to examine the performance of the DBRF method using expression data of all single gene deletions. To this end, a series of gene networks and all single gene deletion mutants for each network are simulated to generate sets of target artificial steady-state gene expression data. After generating the data sets, we apply the DBRF method to these data, and infer a gene regulatory network (Kyoda et al., 2000).

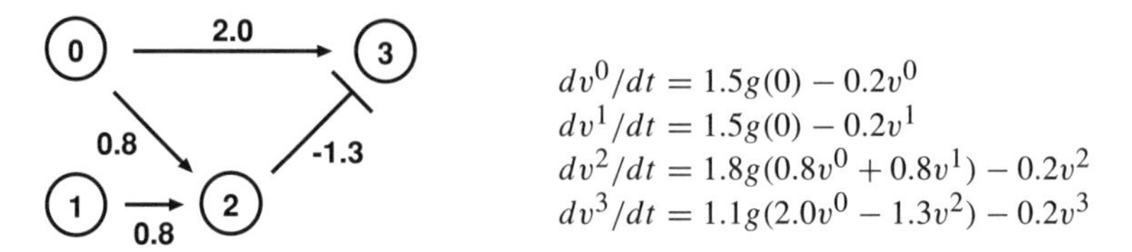

0
2.0
3
0.8
-1.3
1
2
0.8
$dv^0/dt = 1.5g(0) - 0.2v^0$
$dv^1/dt = 1.5g(0) - 0.2v^1$
$dv^2/dt = 1.8g(0.8v^0 + 0.8v^1) - 0.2v^2$
$dv^3/dt = 1.1g(2.0v^0 - 1.3v^2) - 0.2v^3$

(a) A network with weight values (b) The model equations of the network

Figure 3.3 Example of a gene regulatory network model

Network Model

Here, we present the network model used for generating the artificial gene expression data. A gene regulatory network is described as a graph structure consisting of nodes a_n ($n = 0, 1, \cdots, N$), directed edges between nodes with weights, and a function g_n for each node. A node represents a *gene*, and a directed edge represents a gene *regulation*. The weight of a directed edge takes a positive/negative value representing activation/repression effect on the target gene. The expression level of a gene a_n is determined by g_n, which is a nonlinear sigmoidal function reported to describe a gene expression (Mjolsness et al., 1999; Kosman et al., 1998). Thus, the expression level of gene a is described by the following equation:

$$\frac{dv^a}{dt} = R_a g\left(\sum_b W^{ab} v^b + h^a\right) - \lambda_a v^a \tag{3.1}$$

where v^a represents the expression level of gene a, R_a is the maximum rate of synthesis from gene a, and $g(u)$ is a sigmoidal function given by $g(u) = (1/2)[(u/\sqrt{u^2+1}) + 1]$. W^{ab} is a connection-weight matrix element which describes gene regulatory coefficients. $\sum_b W^{ab} v^b$ can be replaced by $\prod_b W^{ab} v^b$, allowing the equation to describe cooperative activation and repression (Mannervik et al., 1999). h^a summarizes the effect of general transcription factors on gene a, and λ_a is a degradation (proteolysis) rate of the product of gene a. We assume that this level always takes a continuous value.

Figure 3.3 shows an example of a small network with four genes. In Figure 3.3(a), each gene a_n is represented by a circle with gene number n. Each directed edge has an effective weight for the target gene. The model equations for each gene are shown in Figure 3.3(b).

The target artificial networks were generated over a range of gene number N and maximum indegree k. For constructing a target network T with N genes and maximum indegree k, the edges were chosen randomly so that the indegree of each gene would be distributed between 1 and k. Besides, each network was generated containing cyclic-regulations, but without containing self-regulations. The parameters in the model equations and regulation type (whether each gene is regulated by gene(s) independently or cooperatively) were randomly determined. For each network, we simulated all single deletion mutants. For each of network sizes N and k, we simulated 100 target networks.

Performance of the DBRF Method

We analyzed the above artificial data with the DBRF method, and compared the inferred networks with the original target networks. The similarity between each inferred network and its target network was evaluated by two criteria, *sensitivity* and *specificity*. Sensitivity is defined as the percentage of edges in the target network that are also present in the inferred network, and specificity is defined as the percentage of edges in the inferred network that are also present in the target network. The results from the experiments over a range of N and k are shown in Table 3.2. As can be seen in Table 3.2, the average of specificity is always higher than that of sensitivity, and sensitivity increases in proportion to the network size N. The average of specificity is about 90% for the indegree $k = 2$, independent of N. The averages of sensitivity and specificity decrease in proportion to the increase of k.

Comparison between Continuous-value Data and Binary Data

One of the major characteristics of the DBRF method is its applicability to continuous-value expression data. To confirm the superiority of using continuous-value expression data, we applied the DBRF method to continuous-value data and binary data, and compared the results. The continuous-value expression data were translated into binary data according to a threshold; the threshold is determined as the middle value be-

Table 3.2 The results from the experiments over a range of N and k. Each measurement is an average over 100 simulated target networks, with standard error given in parentheses.

N	k	Total sim. edges	Total inferred edges	Num. shared edges	*Sensitivity*	*Specificity*
10	2	15.2(1.6)	8.9(3.0)	8.1(2.9)	53.1%	90.4%
20	2	30.5(0.6)	20.6(4.9)	18.6(4.7)	61.1%	90.6%
50	2	75.5(0.7)	60.7(8.7)	54.4(7.9)	72.1%	89.8%
100	2	150.4(0.6)	133.9(10.4)	119.2(9.5)	79.2%	89.1%
20	4	80.0(0.0)	20.3(8.9)	17.0(7.4)	21.3%	84.1%
20	8	117.1(11.8)	15.1(8.6)	9.6(6.1)	8.1%	61.3%

tween the minimum expression level x_{min} (which is zero, because all expression data over single deletion mutant are given) and the maximum expression level x_{max}. The results from the experiments over a range of N and k are shown in Table 3.3. Both sensitivity and specificity, in the case of the continuous-value data, were much higher than those in the case of the binary data.

Table 3.3 The results from continuous-value and binary expression data over a range of N and k. Each measurement is an average over 100 simulated target networks.

N	k	Continuous-value raw data		Binary translated data	
		sensitivity	*specificity*	*sensitivity*	*specificity*
10	2	53.1%	90.4%	20.3%	58.6%
20	2	61.1%	90.6%	20.9%	61.7%
50	2	72.1%	89.8%	22.2%	63.1%
100	2	79.2%	89.1%	22.9%	65.3%
20	4	21.3%	84.1%	9.7%	60.1%
20	8	8.1%	61.3%	6.4%	47.3%

Comparison with the Predictor Method

The predictor method is the most recently reported gene network inference method for steady-state data (Ideker et al., 2000), and thus is considered to be the most powerful method. Therefore, we compared the performance between the DBRF method and the predictor method. The predictor method is designed to analyze binary data, and is not applicable to continuous-value data. Thus, the predictor method was applied to binary data translated from the original continuous-value data as described above, whereas the DBRF method was applied to the original data.

The results show that the performance of the DBRF method is superior to that of the predictor method (Table 3.4). In the case of $N = 20$, $k = 8$, although the sensitivity of the predictor method is slightly higher than that of the DBRF method, the difference is not significant ($P<0.5$). In Table 3.4, the symbol '$*$' means that the predictor method is not applicable because of NP-complete (Ideker et al., 2000). On the other hand, since the cost of the DBRF method is about $O(n^3)$, the DBRF method can infer gene regulatory networks even in the case of $N = 100$, $k = 2$. We also found that the performance of the predictor method is lower than that of the DBRF method even when both methods were applied to the binary translated data (compare Table 3.3, binary translated data and Table 3.4, the predictor method). The performance of the predictor method here was much lower than the previously published results (Ideker et al., 2000). In our experiment, continuous-value gene expression data was generated and then binarized to apply the predictor method, whereas in the previous work, the applied data was directly generated from a Boolean network model (see discussion).

Table 3.4 The results using the DBRF method and the predictor method over a range of N and k. Each measurement is an average over 100 simulated target networks. The symbol '$*$' means computationally infeasible.

N	k	DBRF method		Predictor method	
		sensitivity	*specificity*	*sensitivity*	*specificity*
10	2	53.1%	90.4%	7.1%	12.5%
20	2	61.1%	90.6%	7.9%	8.4%
50	2	72.1%	89.8%	3.3%	7.7%
100	2	79.2%	89.1%	$*$	$*$
20	4	21.3%	84.1%	9.2%	23.4%
20	8	8.1%	61.3%	8.2%	34.8%

APPLICATION TO YEAST GENE EXPRESSION DATA

We applied the DBRF method to a set of yeast gene expression data obtained by DNA microarray (Kyoda et al., unpublished). The gene expression data consisted of gene expression levels of 249 genes in each of single gene deletion mutants for the 249 genes (Hughes et al., 2000). For each gene, expression levels in wild-type and mutants were statistically analyzed, and the significance of difference in gene expression level between wild-type and mutants was evaluated with the p-value (Hughes et al., 2000). In the following experiment, the DBRF method detected the gene expression difference with a p-value of less than 0.01.

Figure 3.4 shows the inferred gene network for the 249 genes. In total, 628 gene regulatory relationships were inferred. It took about 5 seconds to infer this gene regulatory network, suggesting that the DBRF method has scalability for gene network size. The DBRF method will be able to infer a gene regulatory network from a complete set of gene expression data obtained from yeast for all 6,000 single gene deletions, or other higher organisms with a large number of genes.

In the above result, for example, 13 gene regulatory relationships were inferred in MAP kinase signal transduction cascade (Figure 3.5). In these 13 relationships, all 5 regulations from ste12, the only transcription factor in this cascade, were consistent with the known transcriptional regulations (Roberts et al., 2000; Oehlen et al., 1996; Dietzel and Kurjan, 1987; Errede and Ammerer, 1989; Oehlen and Cross, 1998). All 5 regulations to ste12 were coming from genes whose active forms indirectly up-regulate the transcription of ste12, and thus consistent with the known gene regulatory relationships (Roberts et al., 2000; Ren et al., 2000). The remaining 3 relationships were newly discovered in this study. The results suggest

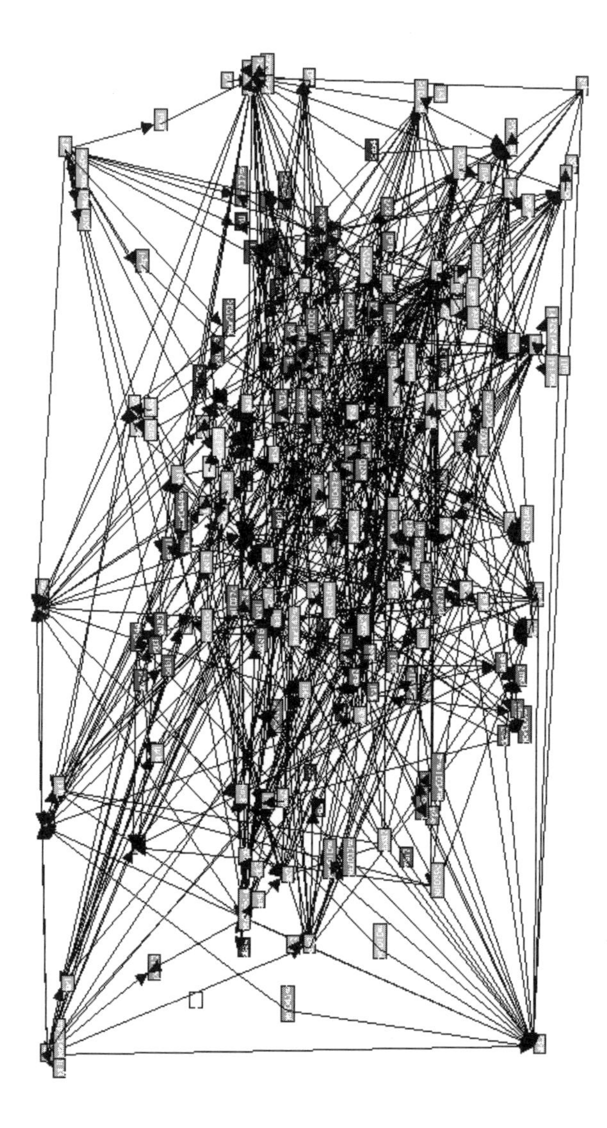

Figure 3.4 The inferred gene network from yeast 249 gene expression data with the DBRF method. Each box represents gene, and each edge represents gene regulatory relationship.

that the DBRF method infers a gene regulatory network quite effectively.

DISCUSSION

In the above sections, we reviewed the algorithm of the DBRF method. The performance of the method and its application to real experimental data were also reviewed. In this section, we discuss several characteristics of this method.

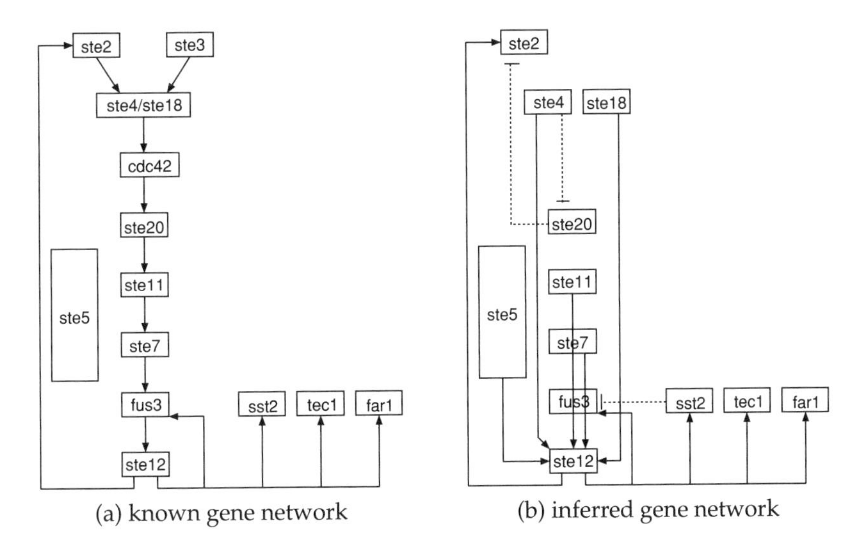

Figure 3.5 MAP kinase signal transduction cascade. (a) Known gene network from recent publications and database. (b) Inferred gene network. Solid-lined arrows mean consistent gene regulations with known gene network, and dashed-lined arrows mean new gene regulations.

Algorithm

The DBRF method infers a gene regulatory network with high specificity but sensitivity is much lower than specificity when the method is applied to a set of single deletion mutant expression data. One reason for the lower sensitivity is that the DBRF method cannot detect the difference between gene expression of wild-type and deletion mutant if a gene is not expressed under the wild-type condition. If we also have the data of overexpression mutants, the sensitivity will increase, because the difference can be detected between wild-type and overexpression mutants. Furthermore, if this happens, the specificity will also increase. The number of inferred direct edges increases with the addition of overexpression data, and those edges prevent several indirect edges from being inferred as direct edges. The expression data monitored over multiple gene mutations can be used for the DBRF method, which will reveal the network structure more accurately. Although the algorithm finds out "indirect" edges and eliminates them, some indirect edges may play a redundant role in the actual network. The DBRF method can distinguish those differences if sufficient experimental data are available.

We found that when using the data of a network which contains cyclic-regulations, a unique network may not necessarily be inferred. We allow gene regulatory networks containing cyclic-regulations in the computational experiments. When such a cyclic network is treated by the DBRF method, the inferred network differs depending on the order of the genes in the interaction matrix I. Several direct edges are eliminated ahead if the eliminating process in Figure 3.2 algorithm starts from an indirect edge. The DBRF method infers several different networks when we change the order of the genes in the interaction matrix I for a given cyclic network. However, all the inferred networks are consistent with the original expression data. It is possible to extend the DBRF method to output all possible candidates of the gene network instead of providing one candidate, but we do not know which is useful for users. This is an open problem when considering cyclic networks.

Advantage of Using Continuous-value Data

Using the continuous-value data was clearly superior to using the binary data. One of the main reasons is that binary data lacks much of the information present in the original data. Binary data cannot reflect the states of even three levels (e.g., increased, non-changed, and decreased level). Assuming three expression levels with intermediate level as wild-type, we can detect the difference of both increased and decreased levels. On the other hand, if we assume two binary levels, the wild-type level should be grouped into either higher or lower level, hence only either level can be detected (e.g., if wild-type level is 1, only lower level can be detected). We think this is a critical disadvantage of applying a Boolean network model, using binary data, to a gene regulatory network.

The DBRF Method versus the Predictor Method

The performance of the DBRF method is much better than that of the predictor method. In our experiment, the sensitivity and the specificity of the predictor method were much lower than the previously published results (Ideker et al., 2000). In our experiment, continuous-value gene expression data were generated, and then the data were binarized before applying the predictor method, whereas in the previous work, the applied data was directly generated from a Boolean network model. We believe that our result is more realistic since real experimental data have continuous values. Unexpectedly, the performance of the DBRF method, even when using binary data, was significantly superior to that of the predictor method. Since the predictor method compares not only between wild-type and mutant but also between different mutants, we expected that the predictor method would be able to infer more candidates of gene regulations than the DBRF method when both methods were applied to the same binary translated data. The DBRF method only compares between wild-type and mutant. We found that the algorithm of minimum set finding, which plays a key role in comparison between different mutants in the predictor method, did not function as intended in this experiment. Data binarization eliminates many relationships between genes, and it naturally reduces the number of truly inferred edges. With those reduced number of true edges, the minimum set finding tends to create many mistakes, thus lowering the sensitivity and the specificity. One of the typical mistakes is to infer edges in the opposite direction.

Since the computational cost of the DBRF method is $O(n^3)$, we can compute large-scale steady-state expression data. On the other hand, the predictor method is computationally infeasible to analyze a large-scale expression data, because the minimum set covering task is NP-complete. Ideker *et al.* suggested that they could solve this problem by setting the maximum number of indegrees k (Ideker et al., 2000). However, this solution is not suitable for real gene regulatory networks. No one knows the maximum number of indegrees, such as the maximum number of transcriptional factors binding to the cis-regulatory region. In the course of examining the predictor method, we also found several cases where the maximum number of indegrees in the network inferred became larger than that in the original target network. This indicates that it is very difficult to set the maximum number of indegrees for the predictor method even if we know this number for real networks.

Application to Yeast Gene Expression Data

Since yeast gene expression data contain noise coming from the experimental procedure, we need to allow slight fluctuation of gene expression to avoid inferring wrong gene regulations. The range of noise for each gene should be determined by statistical analysis of a series of the gene expression data with negative control experiments. The significance of the difference between two data is represented by the p-value calculated with an error model such as the gene-specific error model (Hughes et al., 2000). In the above yeast expression analysis, the DBRF method was allowed to detect only the gene expression difference with significance at $P<0.01$ in the gene-specific error model.

Compared with the known gene network, the DBRF method inferred many indirect gene regulations in MAP kinase cascade. As mentioned above, the DBRF method tends to infer indirect regulations when single gene deletion data are applied. However, in this case, the number of indirect regulations does not decrease even if overexpression data or double mutant data are applied. We found that the indirect gene regulations arise from gene regulations through protein phosphorylation. To infer those post-transcriptional regulations systematically, a large scale protein expression or modification analysis such as protein chip analysis (Zhu et al., 2000) is required. If we obtain protein phosphorylation data, the DBRF method may infer more direct regulations and fewer indirect regulations.

A post-transcriptional regulation cascade regulates the ste12 activity. The DBRF method inferred 5 gene regulations to ste12 coming from genes in this cascade, because the activated ste12 up-regulates its transcription

(Ren et al., 2000). In the following, we consider the case when the DBRF method is applied to DNA microarray data on a set of single deletion mutants. When the activity of a transcription factor is regulated by post-transcriptional regulation cascade, the DBRF method infers gene regulations to the transcription factor coming from genes in this cascade if the transcription factor self-regulates its transcription. Alternatively, gene regulations from those genes to the direct targets of the transcription factor are inferred if the self-regulation does not exist. Hence, althouogh the DBRF method basically infers transcriptional regulation between genes, it also provides valuable information on post-transcriptional regulations.

CONCLUSION

In this chapter, we reviewed the DBRF method, a method for inferring a gene regulatory network from steady-state gene expression data. The DBRF method is applicable to continuous values of expression data, whereas the other methods that also use steady-state data can only deal with binary data. The performance of the DBRF method was evaluated by varying the network size, indegree of each gene, and the data characteristics (continuous-value or binary). The DBRF method was shown to be superior to the other existing methods. We also reviewed the applicability of the DBRF method to real gene expression data. The method was shown to have scalability for large-scale gene expression data. The accuracy of the method was shown in MAP kinase cascade, where several consistent and new relationships were inferred. Overall, the DBRF method will be a powerful tool for genome-wide gene network analysis.

References

Akutsu, T., Kuhara, S., Maruyama, O., and Miyano, S. (1998). Identification of gene regulatory networks by strategic gene disruptions and gene overexpressions. *Proc. 9th ACM-SIAM Symp. Discrete Algorithms* pp.695–702.

Dietzel, C. and Kurjan, J. (1987). Pheromonal regulation and sequence of the Saccharomyces cerevisiae SST2 gene: a model for desensitization to pheromone. *Mol. Cell. Biol.* 7(12):4169–4177.

Errede, B., and Ammerer, G. (1989). STE12, a protein involved in cell-type-specific transcription and signal transduction in yeast, is a part of protein-DNA complexes. *Genes and Dev.* 3(9):1349–1361.

Gengyo-Ando, K., and Mitani, S. (2000). Characterization of mutations induced by ethyl methanesulfonate, UV, and trimethylpsoralen in the nematode Caenorhabditis elegans. *Biochem. Biophys. Res. Comm.* 269(1):64–69.

Gross, J., and Yellen, J. (1999). *Graph theory and its applications.* CRC Press.

Hughes, T.R., Marton, M.J., Jones, A.R., Roberts, C.J., Stoughton, R., Armour, C.D., Bennett, H.A., Coffey, E., Dai, H., He, Y.D., Kidd, M.J., King, A.M., Meyer, M.R., Slade, D., Lum, P.Y., Stepaniants, S.B., Shoemaker, D.D., Gachotte, D., Chakraburtty, K., Simon, J., Bard, M., and Friend, S.H. (2000). Functional discovery via a compendium of expression profiles. *Cell,* 102:109–126.

Ideker, T.E., Thorsson, V., and Karp, R.M. (2000). Discovery of regulatory interactions through perturbation: inference and experimental design. *Proc. Pacific Symp. Biocomputing 2000* pp.305–316.

Kosman, D., Reinitz, J., Sharp, D.H. (1998). Automated assay of gene expression at cellular resolution. *Proc. Pacific Symp. Biocomputing '98* pp.6–17.

Kyoda, K.M., Morohashi, M., Onami, S., and Kitano, H. (2000). A gene network inference method from continuous-value gene expression data of wild-type and mutants. *Genome Informatics* 11:196–204.

Kyoda, K., Morohashi, M., Onami, S., and Kitano, H. (2001). Unpublished data.

Liang, S., Fuhrman, S., and Somogyi, R. (1998). REVEAL: a general reverse engineering algorithm for inference of genetic network. *Proc. Pacific Symp. Biocomputing '98* pp.18–29.

Mannervik, M., Nibu, Y., Zhang, H., and Levine, M. (1999). Transcriptional coregulators in development. *Science* 284(5414):606–609.

Mjolsness, E., and Mann, T., Castaño, R., and Wold, B. (1999). From coexpression to coregulation: an approach to inferring transcriptional regulation among gene classes from large-scale expression data. *Tech. Rept. JPL-ICTR-99-4*, Jet Propulsion Lab. NASA.

Morohashi, M. and Kitano, H., (1999). Identifying gene regulatory networks from time series expression data by *in silico* sampling and screening. *Proc. 5th Euro. Conf. Artificial Life* pp.477–486.

Oehlen, L.J., McKinney, J.D., and Cross, F.R. (1996). Ste12 and Mcm1 regulate cell cycle-dependent transcription of *FAR1*. *Mol. Cell. Biol.* 16(6):2830-2837.

Oehlen, L., and Cross, F.R. (1998). The mating factor response pathway regulates transcription of TEC1, a gene involved in pseudohyphal differentiation of Saccharomyces cerevisiae. *FEBS Lett.* 429(1):83–88.

Ren, B., Robert, F., Wyrick, J.J., Aparicio, O., Jennings, E.G., Simon, I., Zeitlinger, J., Schreiber, J., Hannett, N., Kanin, E., Volkert, T.L., Wilson, C.J., Bell, S.P., and Young, R.A. (2000). Genome-wide location and function of DNA binding proteins. *Science* 290(5500):2306–2309.

Roberts, C.J., Nelson, B., Marton, M.J., Stoughton, R., Meyer, M.R., Bennet, H.A., He, Y.D., Dai, H., Walker, W.L., Hughes, T.R., Tyers, M., Boone, C., and Friend, S.H. (2000). Signaling and circuitry of multiple MAPK pathways revealed by a matrix of global gene expression profiles. *Science* 287(5454):873–880.

Sharp, P.A. (1999). RNAi and double-strand RNA, *Genes Dev.* 13(2):139–141.

Spradling, A.C., Stern, D., Beaton, A., Rhem, E.J., Laverty, T., Mozden, N., Misra, S., and Rubin, G.M. (1999). The Berkeley Drosophila Genome Project gene disruption project: single P-element insertions mutating 25% of vital Drosophila genes, *Genetics* 153(1):135–177.

Thieffry, D., and Thomas, R. (1998). Qualitative analysis of gene networks. *Proc. Pacific Symp. Biocomputing '98* pp.77–88.

Winzeler, E.A., Shoemaker, D.D., Astromoff, A., Liang, H, Andeson, K., Andre, B., Bangham, R., Benito, R., Boeke, J.D., Bussey, H., Chu, A.M., Connelly, C., Davis, K., Dietrich, F., Dow, S.W., Bakkoury, M.E., Foury, F., Friend, S.H., Gentalen, E., Giaever, G., Hegemann, J.H., Jones, T., Laub, M., Liao, H., Liebundguth, N., Lockhart, D.J., Lucau-Danila, A., Lussier, M., M'Rabet, N., Menard, P., Mittmann, M., Pai, C., Rebischung, C., Revuelta, J.L., Riles, L., Roberts, C.J., Ross-MacDonald, P., Scherens, B., Snyder, M., Sookhai-Mahadeo, S., Storms, R.K., Véronneau, S., Voet, M., Vockaert, G., Ward, T.R., Wysocki, R., Yen, G.S., Yu, K., Zimmermann, K., Philippsen, P., Johnston, M., and Davis, R.W. (1999). Functional characterization of the *S. cerevisiae* genome by gene deletion and parallel analysis. *Science* 285(5429):901–906.

Zhu, H., Klemic, J.F., Chang, S., Bertone, P., Casamayor, A., Klemic, K.G., Smith, D., Gerstein, M., Reed, M.A., and Snyder, M. (2000). Analysis of yeast protein kinases using protein chips *Nature Genetics* 26(3):283–289.

4 The Analysis of Cancer Associated Gene Expression Matrices

Mattias Wahde and Zoltan Szallasi

An important part of the analysis of cancer associated gene expression matrices is the identification of a subset of genes displaying consistent mis–regulation in a given type of tumor samples. Such a subset of genes forms, together with an appropriate function, a separator that can distinguish between normal and tumor samples. The identification of separators is a difficult problem, due to the very large sizes of the search spaces involved. In this paper, we introduce and discuss briefly a method for identification of separators using genetic algorithms.

Due to the high level of gene expression diversity detected in cancer, separators can appear by chance. In order to find the true separators, it is important to weed out such chance separators. There are several statistical methods for estimating whether the appearance of a given separator is due to chance. The accuracy of such tests will, however, depend on the null hypothesis provided by the data structure. In this paper we introduce and describe generative models that simulate random, discrete gene expression matrices which retain the key features of massively parallel measurements in cancer. These include the number of changeable genes and the level of gene co–regulation as reflected in their pair–wise mutual information content. By analyzing several cancer–related data sets, we demonstrate that the probability of the chance appearance of separators can be underestimated by many orders of magnitude if random and independent selection of mis–regulated genes is assumed, instead of using more advanced generative models as outlined in this paper.

INTRODUCTION

The recent publication of several cancer associated large-scale gene expression matrices has clearly indicated that tumor biology has entered a new phase of analytical approaches. These matrices contain quantitative information about a large number of directly measured parameters, usually gene expression levels, that are typically listed as the rows of the matrix. The columns in these experiments correspond to different phenotypes such as different types of tumors or different treatments of either normal or neoplastic cells.

There are two obvious ways of exploiting cancer associated gene expression matrices. Identification of separators or gene expression functions (Szallasi, 1998) determines a subset of genes the status of which, when coupled by an appropriate rule, will define the phenotypic state of cells. The classification of phenotypic samples on the other hand is supposed to identify subsets of samples with above average molecular similarity. These subsets can be later used to search for common genetic markers. This procedure, which was recently termed as tumor class discovery in cancer research (Golub et al., 1999), is supposed to yield a group of tumor samples sharing a common set of genetic markers. In principle, these two types of analysis are overlapping since a new tumor subclass is supposed to be determined by a subset of genes that obviously form a separator differentiating the new phenotype from the rest of the samples. In practice, however, the two methods show a clear distinction regarding the possible number of genes involved. Identification of separators searches for the fewest possible genes that will distinguish between phenotypes, whereas classification or cluster analysis may be based on a much larger subset, hundreds or even thousands of genes.

Initial efforts in the field have met with limited success, which could be the result of at least three possible causes: (a) working with incomplete or inaccurate data sets (i.e. some of the relevant genes were not measured), (b) working with analytical tools of inadequate power, and (c) ignoring the special characteristics of massively parallel gene expression matrices. When identifying separators the last two of the possible difficulties manifest themselves in the following way: we need to be able to extract correct separators within a reasonably short period of time and then show that these separators did not emerge by accident in the gene expression matrix.

Here we will address both difficulties: first we show that the power of identifying separators could be significantly improved by genetic algorithms and then demonstrate the importance of internal data structure in statistically validating separators that were extracted from cancer associated gene expression matrices. The successful identification of separators or tumor subclasses will depend on several factors including (1) the number of genes involved; (2) the complexity of the rule between these genes; (3) the available number and diversity of gene expression samples for a given phenotype; (4) the overall diversity of gene expression patterns between normal and tumor samples; and (5) the overall noise level of gene expression measurements.

A rather unfavorable, but not that unlikely scenario holds that the number of sufficiently diverse gene expression patterns will be too limited relative to the order (number of genes involved) and complexity of separators involved in cancer. In this case our analytical efforts will probably fail unless we understand the overall mechanism of how gene expression patterns are generated in cancer. For example, it is obvious that not all gene expression patterns are compatible with life. This may allow excluding significant portions of the gene expression space from further consideration, which would significantly improve our analytical chances.

Statistical analysis of cancer associated gene expression matrices also emphasizes the importance of understanding internal data structure. Cancer associated gene expression patterns show a high level of diversity. The average number of mis-regulated genes is on the order of 10% of all genes expressed in the given cell type (Perou et al., 1999) which inevitably leads to the accidental appearance of separators and clusters in these data sets. However, the highly diverse differential gene expression patterns in cancer are the result of transitions of a self-consistent genetic network contained within the cell (Klus et al., 2001). This involves the co-regulation of genes that is reflected by the internal data structure of cancer associated gene expression matrices. Ignoring the co-regulation of genes may lead to a significant misestimate of statistical significance. We will overcome this problem by introducing generative models in order to estimate the probability of accidental features of cancer associated gene expression data sets.

SEPARATORS

The purpose of separators is to identify patterns of gene expression indicative of neoplasticity. Thus, a separator $S = S(g_1, g_2, \ldots, g_K)$ is a discrete function of several inputs which takes the value 1 if the corresponding sample is in a neoplastic state and 0 otherwise. In this paper, we will focus on the problem of identifying separators in discretized gene expression data. Continuous cDNA microarray measurements can be converted into ternary data as described by (Chen et al., 1997). Their algorithm first calibrates the data internally to each microarray and statistically determines whether the data justifies the conclusion that a given gene is up- or down-regulated at a certain confidence level. Accordingly, in these data sets, the expression level of each gene can take one of three values, namely -1 (down–regulated), 0 (unchanged), or 1 (up–regulated).

Cancer associated gene expression measurements have provided two main types of data sets so far. In the first case, all samples are in the neoplastic state (i.e. $S = 1$), and the down– or up–regulation is measured relative to an appropriate normal control. In the second case, the data set consists of both neoplastic ($S = 1$) and normal tissue samples ($S = 0$). In such cases the up– or down–regulation of a gene can, for instance, be defined relative to the average expression level of that gene throughout the normal samples.

Let us here consider the case in which all samples are in the neoplastic state $S = 1$, and let N denote the number of genes in each sample, M_- and M_+ the number of down– and up–regulated genes, respectively, and M their sum, i.e. $M = M_- + M_+$. The number of samples is denoted E. The easiest case is a one–gene separator ($K = 1$) when all tumor samples carry at least one consistent gene mis-regulation. We provided a detailed statistical analysis of single gene separators elsewhere (Wahde et al., 2001). Clearly, any set of genes $(g_1, ..., g_K)$ for which there exists at least one sample such that $g_1 = g_2 = ... = g_K = 0$ *cannot* describe a separator, since some change in the expression levels is needed to arrive at the neoplastic state. Thus, in this case, the first step in identifying a separator of K inputs, is to find all combinations of K genes such that, in each sample, at least one of the K genes is down– or up–regulated. Any such combination of genes defines a separator.

IDENTIFICATION OF SEPARATORS IN NOISY DATA

The discussion above concerning separators is somewhat simplistic in that it assumes data to be more or less noise–free, so that a simple deterministic Boolean function of a few genes can separate tumors from normal samples. Given the high noise levels in gene expression data, and the great diversity of gene expression in cancer samples, a more realistic approach would instead make the assumption that genes related to cancer are often, but not always, mis–regulated, so that a given degree of mis–regulation of a weighted *average* of a few such genes could be used as an indicator of a tumor. Several such indicators could then be combined using e.g. Boolean functions, much in the same way as the expression levels of single genes were combined in the method described above.

The problem, of course, is to identify the relevant genes. As an example, consider a case where a weighted average of the expression level of four genes is to be used as an indicator. A typical sample includes measurements of thousands of genes. With $10^{3.5}$ measured genes, an exhaustive search would require checking of order 10^{12} different combinations.

Clearly, exhaustive searches are not optimal in these cases. A good alternative is provided by genetic algorithms, which we will now introduce.

Genetic algorithms

Genetic algorithms (GAs) are based on the principles of Darwinian evolution, involving gradual hereditary change of candidate solutions.

When a GA is applied to a problem, a population (i.e. a set) of candidate solutions is maintained. The candidate solutions themselves are called individuals. The information needed to form the individuals is encoded in strings of digits known as chromosomes, where, in keeping with the biological terminology, the individuals digits are referred to as genes.

A GA is initialized by assigning random values to the genes in the chromosomes. The next step is to decode the chromosomes of each individual, and to perform the evaluation. Both the decoding process and the evaluation procedure are, of course, problem dependent. As a specific example, we may consider a case in which the chromosomes encode the identities of N genes that are to be used for distinguishing tumors from normal samples, by forming an average over the expression levels of the N genes, and comparing it to a threshold which is also encoded in the chromosome. In such a case, the decoding procedure simply identifies the genes whose expression levels are to be averaged. Then, the average of those genes are formed, for each sample, and the result is compared with the threshold obtained from the chromosome. If the average exceeds the threshold, the sample is placed in the tumor category, otherwise it is placed in the normal category. When all samples have been evaluated, a performance measure (the fitness) is assigned to the individual. In the classification task, the fitness can, for instance, be defined as the fraction of samples that are correctly classified.

When all the individuals of the first generation have been evaluated, the second generation is formed by selecting individuals in a fitness-proportionate way, i.e. such that individuals with high fitness have a larger probability of being selected than individuals with low fitness. When two individuals have been selected, their offspring is formed through crossover (with a certain probability, usually taken to be close to 1) and mutation. In crossover, the chromosomes of the two individuals are cut at a randomly chosen point (the crossover point), and the first part of the first chromosome is combined with the second part of the second chromosome, and vice versa to form two new chromosomes. Each of the two new chromosomes is then subjected to mutation, during which a random number is drawn for each gene. If the random number is smaller than a pre–specified mutation probability, then the gene is assigned a new, random value. If not, the gene is left unchanged.

The procedures of selection, crossover, and mutation is repeated until there are as many new chromosomes as in the first generation. Then, the old chromosomes are replaced by the new ones, and all the individuals of the new generation are evaluated. The third generation is formed by selecting individuals from the second generation, and performing crossover and mutation on their chromosomes. The whole procedure is repeated for a large number of generations, until a satisfactory solution to the problem has been achieved.

There exists many different versions of genetic algorithms that use different ways of selecting individuals for reproduction, different ways of assigning fitness values, as well as different methods of maintaining the population. Furthermore, in practical applications of GAs, the procedure often becomes more complex than in the simple case introduced above. For example, it is not uncommon that chromosomes of varying length need to be used, which makes the crossover procedure more complicated. For a more comprehensive introduction to GAs, which also discusses some of the more advanced issues, see e.g. (Mitchell, 1995).

Figure 4.1 A simple chromosome for gene identification.

A method for automatic identification of genes

We will now briefly discuss the issue of identification of cancer relevant genes using GAs, and suggest a specific method for this problem.

The first step in the application of a GA is to select an encoding scheme for the chromosomes. Clearly, it is not difficult to devise chromosomes that encode very complex classifiers. However, a good classifier should preferably be as simple as possible, especially in cases where the number of available samples is strongly limited. A classifier which uses, say, three parameters and which can classify perfectly a set of 50 samples is obviously more likely to be reliable than a classifier that uses, say, 25 parameters. On the other hand, it is difficult to know beforehand what the appropriate number of parameters is, and so one should preferably allow it to be determined by the GA.

For simplicity, let us consider a specific method in which a simple average (i.e. with equal weights for all genes) is used, and where there is only one condition. In this case, all that needs to be encoded are the genes from which the average should be formed, and the value of the threshold which determines the category into which an evaluated sample will fall. Note that, in order not to confuse the genes of the expression matrix with the genes of the chromosomes used by the GA, we will in this section refer to the latter as "entries" rather than genes.

Here, an encoding scheme of the kind shown in Figure 4.1 can be used. The first entry of the chromosome encodes the number (M) of genes for which the average should be formed, and the following M entries determine the identity of those genes. The final entry of the chromosome set is the value of the threshold. The entries take value between 0 and 1, and are then rescaled to form numbers between 1 and N_g (for the entries identifying genes) and between -1 and 1 for the final entry of the chromosome.

Normally, the chromosomes of the first generation are generated randomly. However, when there is biological information present, it should of course be used. An inspection of a gene expression matrix indicates that some genes are less likely to be of importance than others. For instance, some genes never change (in any sample), whereas others change in a more or less random way.

Thus, a sensible approach is to first rank the genes in descending order of importance, placing at the top those genes that show a large and consistent difference between the tumor samples and the normal samples. The measure of such a difference can be selected in various ways: mutual information is one possibility (Butte and Kohane, 2000). An alternative relevance measure can be defined as

$$r = (|\frac{n_{-}^{(1)}}{n_{(1)}} - \frac{n_{-}^{(2)}}{n_{(2)}}| + |\frac{n_{0}^{(1)}}{n_{(1)}} - \frac{n_{0}^{(2)}}{n_{(2)}}| + |\frac{n_{+}^{(1)}}{n_{(1)}} - \frac{n_{+}^{(2)}}{n_{(2)}}|)\max(n_{-}^{(2)}, n_{+}^{(2})), \quad (4.1)$$

where superscript 1 denotes the normal samples and superscript 2 the tumors, and where n_{-}, n_0, and n_{+} denote the number samples in which the gene in question is down-regulated, unchanged, and up-regulated, respectively. This measure assigns high relevance values to genes that show a large difference in the expression patterns between normal samples and tumors, and for which the mis-regulation of the tumor samples is consistent, i.e. either down-regulation or up-regulation.

Now, in order to begin the search starting from the simplest possible classifiers, the chromosomes of the initial population should all have $M = 1$ (corresponding to a simple $K = 1$ separator as defined in the beginning of this chapter) and the decoding scheme identifying the genes should be based on the relevance ranking described above. Thus, the entry defining the single gene should be initialized to a low value. As an example, if it were to be set exactly to 0, the corresponding gene would be the one with the highest score in the relevance ranking.

A useful fitness measure in this case would be

$$f = max(p - 1, pq^{(M-1)}), \quad (4.2)$$

where p denotes the number of correctly classified samples, q is a number slightly smaller than 1, and M, as usual, denotes the number of entries identifying genes in the chromosome of the classifier. Thus, an optimal classifier would be one that could correctly classify all samples using the measurement of only one gene. The max function is needed in order to prevent that a classifier with a lower value of p receives a higher fitness value than one with a higher value of p. Thus, a limit on the punishment for overly complicated classifiers is introduced.

This concludes our brief description of a possible GA-based method for the identification of interesting genes in neoplastic samples. Needless to say, the procedure could be improved in various ways, for instance by allowing Boolean combinations of several conditions. Such improvements will not be discussed here, however.

Furthermore, not all samples should be used in the determination of the classifier: some should be retained for validation purposes. If the number of samples is small, a procedure can be implemented in which a random set of samples is used by the GA, and the rest are used for the validation. Several runs can be carried out, with different validation sets. Those genes that appear in many or all of the classifiers obtained by the

GA are them likely to be interesting candidates for further study.

Statistical validation of separators extracted from gene expression matrices

In the previous section we have presented search strategies for separators in gene expression matrices. The high level of gene expression diversity in cancer samples, however, makes it probable that separators can occur by chance. In order to identify the true separators in a data set, such chance separators must first be identified and removed. In order to do so, one needs some way of estimating the probability that any given separator is due to chance. This probability can be readily estimated by analytical means only in the case of low order separators (preferably $K = 1$) and special gene expression matrices, when mis-regulated genes are randomly and independently selected (Wahde et al., 2001). In more realistic data sets analytical calculations become intractable and one needs to rely on computer simulations. In other words, one needs a *generative model* which can generate artificial data sets, the analysis of which can provide estimate of the probability of the chance appearance of a separator.

Generative models

The aim of a generative model is to produce an artificial and random data matrix which shares the essential characteristics of the original data matrix. The artificial data obtained by means of the generative model can then be used to form null hypotheses for the estimation of the probability of separators discovered in the real data set, thus making it possible to distinguish chance separators from actual separators. Generative models can be derived from either theoretical considerations or empirical observations. In cancer research, theory–based generative models can use either genetic network modeling or aneuploidy driven gene mis–regulation as their starting point.

Malignant transformation can be considered as an attractor transition of a self–organizing gene network (Kauffman, 1993; Szallasi and Liang, 1998) providing numerical estimates about the overall quantitative features of attractor transition like the expected number of up– or down–regulated (with a common term, mis–regulated) genes. There is an increasing evidence of the ploidy regulation of gene expression levels as well (Galitski et al., 1999). We have provided initial indications that the aneuploidic distribution of chromosomes may also be used to model the expected gene expression patterns in cancer (Klus et al., 2001). At the current stage of theory and available data sets, however, we can best rely on generative models based on empirical observations. This approach starts with extracting overall quantitative features of cancer associated gene expression matrices. These include the number of genes that can be mis–

regulated, the ratio of up– versus down–regulated genes and the level of co–regulation of mis–regulated gene groups. We will now discuss two very different approaches to generative models which will shed some light on the importance of a careful selection of such models.

The first method simply forms a randomized gene expression matrix while preserving certain overall features of the real data matrix, such as the number of mis–regulated genes in each sample. Mutual information based generative models, which is the second method introduced here, preserve additional features of the real data, namely the co–regulation of genes.

Randomization based generative models

The simplest method of generating artificial data consists simply of inserting, for each sample, M_+ 1's and M_- -1's randomly in a null $N \times E$ matrix. In general, the values of M_- and M_+ will of course vary from sample to sample, so either an average value or the actual values of M^i_- and M^i_+ $(i = 1, \ldots, E)$ from the real data can be used. It turns out that the formula for the expected number of separators is very sensitive to the values of M^i_- and M^i_+, and therefore the use of average values is not to be recommended. The randomization method that uses the actual values of M^i_- and M^i_+, will be referred to as *simple randomization*. As a simple example, consider the case of a $K = 2$ gene separator. Assume that two genes, denoted g_1 and g_2, are being studied. In a given sample i, the approximate probability $p^i_s(2)$ of at least one of these two genes being changed (up– or down–regulated) is

$$p^i_s(2) = 1 - (p^i_0)^2, \tag{4.3}$$

where $p^i_0 = (N - M^i)/N$ denotes the probability of a given gene being unchanged ($M^i = M^i_+ + M^i_-$, where M^i_+ and M^i_- denote, as before, the number of up– and down–regulated genes in sample i, respectively). Note that the approximation is valid as long as $1 << M^i << N$. In a typical neoplastic sample it is safe to make this assumption, since $\sim$10% of the genes are changed (i.e. $M^i \sim 0.1N$). The probability of at least one of the genes being changed in each of the E samples equals

$$P_s(2) = \prod_{i=1}^{E} p^i_s(2) \equiv \prod_{i=1}^{E} (1 - (p^i_0)^2). \tag{4.4}$$

Thus, the expected number of such separators is

$$N_s(2) = \binom{N}{2} P_s(2). \tag{4.5}$$

Table 4.1 A $K = 2$ separator. The final column shows the value of the function S (the separator) for the given input configuration.

g_1	g_2	S
-1	-1	1
-1	0	0
-1	1	0
0	-1	1
0	0	0
0	1	0
1	-1	0
1	0	0
1	1	1

Generalizing these formulae, it is easy to see that the expected number of separators of K inputs is

$$N_s(K) = \binom{N}{K} P_s(K) \equiv \binom{N}{K} \prod_{i=1}^{E} p_s^i(K) \tag{4.6}$$

where

$$p_s^i(K) = 1 - (p_0^i)^K. \tag{4.7}$$

This analysis gives an estimate of the *total* number of separators of K inputs expected in a randomized artificial data set. Using similar methods, the approximate probability of discovering any specific separator in artificial data can also be obtained. In the case of K inputs, the total number of combinations of the input variables equals 3^K. The estimate of the probability of a specific separator begins by the computation of the probability, for one sample i, of obtaining one of those combinations for which $S = 1$. This probability is denoted p_R^i. The expected number of separators in the data set is then given by

$$N_R = \binom{N}{K} P_R \equiv \binom{N}{K} \prod_{i=1}^{E} p_R^i. \tag{4.8}$$

As an example, consider a separator defined by the entries of Table 4.1. In any given sample i, the probability of having $S = 1$ equals

$$p_R^i = p_{-1,-1}^i + p_{0,-1}^i + p_{1,1}^i = (p_-^i)^2 + p_0^i p_-^i + p_+^i p_+^i, \tag{4.9}$$

where $p_0^i = (N - M_-^i - M_+^i)/N$, $p_-^i = M_-^i/N$, and $p_+^i = M_+^i/N$. The approximate number of expected separators of this type is then

$$N_R = \binom{N}{2} P_R = \binom{N}{2} \prod_{i=1}^{E} p_R^i. \tag{4.10}$$

Armed with the tools presented above, we can proceed to analyze sets of cancer–related gene expression data. As an example, we have analyzed the colon cancer data published by Alon et al. (1999). These data consist of DNA-oligomer chip based gene expression measurements on 2,000 genes in 22 patient matched neoplastic and normal samples.

According to Eq. 4.6, increasing the sample number (in this case to 22) decreases the expected number of separators appearing by chance. Indeed, applying this equation, the expected number of separators with $K = 2$, assuming random and independent selection is found to be $2.3 \times 10^{-12} <<< 1$. On the other hand, the actual number of separators with $K = 2$ was equal to 1 for this data set, strongly suggesting that this separator might play a role in colon cancer.

Analysis of other cancer–related data sets yield essentially identical result: the number of separators found in the real data set always exceeds by far the number of separators expected on the base of random and independent selection.

The assertion that these results are significant hinges on one important assumption, namely that random and independent selection forms a reasonable null hypothesis for cancer–related data. Alas, this is not the case: the data structure of cancer–related gene expression is in fact very far from that obtained using random and independent selection. This is readily noticed when the distribution of pair-wise correlation of gene expression changes is examined. For discretized data mutual information is an appropriate measure, which is high (close to its maximum, 1) if the mis-regulation of a given gene is highly indicative of the mis-regulation of another gene. (For details see e.g. (Klus et al., 2001; Butte and Kohane, 2000)) In cancer associated gene expression matrices the distribution of *pairwise mutual information* indicates a far from independent selection of mis-regulated genes.

In order to illustrate this we have performed an analysis on the breast cancer associated gene expression matrix published by Perou et al. (1999). This publicly available data set contains cDNA microarray based relative expression levels of about 5,600 genes for a number of both normal and neoplastic breast epithelial samples. For our analysis we have used only gene expression measurements derived from either breast cancer cell lines or primary breast tumors, 16 samples altogether. We have retained only those genes in our analysis that showed an at least 3.5–fold up– or down–regulation in at least two samples.

Using these threshold values we have transformed the original data set into a 1082x16 ternary data matrix. The upper panel of Figure 4.2 shows a histogram of the pairwise mutual information distribution obtained from the real data, whereas the middle panel shows the mutual information distribution obtained from a data set using simple randomization of the data matrix. Clearly, the two distributions are very different, which indicates that the assumption of random and independent selection does not form a reliable null hypothesis for cancer–related gene expression data.

Mutual–information based generative models

As a first step towards the identification of a better null hypothesis, one should take into account two restrictions present in biological systems. First, not every gene can be mis–regulated. The number of changeable genes can be calculated as described elsewhere (Wahde et al., 2001) by conditional probabilities. Second, mis–regulated genes are not independently selected. Gene expression levels in cancer are determined by several factors, such as the regulatory input of other genes and the actual DNA–copy number of the given gene present in a cell (Galitski et al., 1999). This will obviously lead to a high level of interdependence between gene expression levels which is readily quantified by mutual information content.

The aim of mutual information based generative models is to produce random gene expression matrices while retaining the overall level and distribution of co–regulation of mis–regulated genes. There are several possible algorithms to achieve this. Here we present a strategy that starts with a random rearrangement of a real data matrix and then a simple form of evolutionary algorithm keeps rearranging this randomized matrix until the new mutual information distribution will closely approximate the distribution detected in the real data.

Generative algorithm

The generative algorithm begins by generating a random data matrix R, by rearranging the matrix elements of the real data set D. A simple algorithm for arriving at a data set of this type is defined as follows: Loop through all genes. For each gene, loop through each sample, select randomly another sample, and swap the corresponding matrix elements. Note that, with this procedure, the values of M_- and M_+ will change, since they are measured column-wise. However, since the computation of mutual information is based on comparison of genes (rows in the expression matrix), rather than samples (columns) this is the correct way to randomize the matrix in this case. This randomization method will be referred to as **permutative randomization**. Once the permutative randomization has been performed a histogram of pairwise mutual information values is generated. A similar histogram is also generated for the real data set, and the distance between the two histograms is computed as

$$\Delta(H_G, H_D) = \sum_{m=1}^{N_{\text{bins}}} \frac{|H_G(m) - H_D(m)|}{\max(H_D(m), 1)}, \tag{4.11}$$

where H_G is the histogram for the generated data set, H_D is the histogram for the real data set, and N_{bins} is the number of bins in the histograms, for which the bin width thus equals $1/N_{\text{bins}}$. The algorithm then proceeds as follows: A gene j is selected at random among the N genes, and its contribution to the histogram is computed by checking the pairwise mutual information between gene j and all other genes. The contribution of gene j to the histogram is subtracted, and the matrix elements in the corresponding row of the data matrix are rearranged, with probability p_{swap} by the same swapping procedure as was used in the permutative randomization algorithm. Then, the new contribution of gene j to the histogram is computed and the histogram thus obtained is compared with the histogram present before the rearrangement of gene j.

If the distance is smaller than before the rearrangement, the new histogram (and, of course, the corresponding matrix) is kept. If not, the old matrix, and the old histogram, are retained and thus only improvements are kept. This procedure – selection of a random gene, subtraction from the histogram, partial rearrangement, and formation of the new histogram, and finally selection of either the old or the new configuration – is repeated many times, until the distance between the histogram for the artificial data and that of the actual data is smaller than user–defined critical value Δ_c. Usually, Δ_c was taken to be of order 10% of the initial distance between D and R.

This algorithm, which is relatively easy to implement, will restore the original distribution of gene co-regulation in a random fashion as demonstrated in the lower panel of Figure 4.2. In addition to the breast cancer data set introduced above we have examined the effect of mutual information based generative models on two other gene expression matrices derived from various cancer types. This analysis has yielded some insightful results. We were focusing on the chance appearance of $K = 2$ separators. Applying Eq. (4.6) we found for the breast cancer data set, that the expected number of separators assuming random and independent selection is 8.6. This was also confirmed by our simple randomization generative model, yielding an estimate of 8.5 ± 7.7 separators. However, the actual number of potential separators, obtained from the real data set, equals 16,997. Clearly, a comparison with the randomized data matrix would indicate that this is a very significant number indeed. Comparing, however, with the results obtained using the generative algorithm, the result is very different. In this case, the average number of expected separators equals $25,481 \pm 897$. We have also analyzed the gene expression data published by Khan et al. (1998). This data set consisted of 13 samples altogether, seven of them alveolar rhabdomyosarcoma samples and the rest commonly used human cancer cell lines. The data matrix contained ternary expression information about 1248 genes. The actual number of separators for this data set was 16,124. Using Eq. (4.6), an estimate of $0.017 << 1$ separators was obtained, again much lower than the actual value. Using instead the generative algorithm, an average of $17{,}252 \pm 133$ separators were obtained. Finally, we have analyzed a colon cancer associated gene expression matrix published by Alon et al. (1999).

DNA-oligomer chip based gene expression measurements were published on 2000 genes in twenty two patient matched neoplastic and normal colon samples by Alon et al. (1999). According to Eq. (4.6), increasing the sample number (in this case to 22) decreases the expected number of separators appearing by chance. Indeed, applying this equation, the expected number of separators assuming random and independent selection is found to be $2.3 \times 10^{-12} <<< 1$. On the other hand, the actual number of potential separators with $K = 2$ was equal to 1 for this data set, suggesting that this separator might play a role in colon cancer. This assumption, however, must be reevaluated after applying the mutual information based generative models. If the essential structure of the colon cancer associated gene expression matrix is retained then the expected number of separators is increased by twelve orders of magnitude to 3.7 ± 1.4. This result has questioned the significance of the potential separator found in the data. This doubt was reinforced by the fact that neither gene involved in the separator has any documented involvement with any forms of human cancer. These initial results indicated, that the chance of random appearance of separators is severely underestimated by ignoring the high level of co-regulation of mis-regulated genes.

Nevertheless, it was also clear that for more accurate estimates our algorithm needs further refinement. The distribution presented in the lower panel of Figure 4.2 showed signs of over-fitting by accurately reproducing the original distribution instead of approximating a theoretical curve describing the mutual information distribution of cancer associated gene expression matrices. This over-fitting also manifested itself in our numerical results. For each of the three data sets the number of expected separators provided by the generative algorithm was 1.07 to 3.7-fold higher than the number of separators present in the actual data sets. On average, however, these gene expression matrices should contain at least as many separators as expected by chance. This over-fitting can probably be avoided by fitting a theoretical curve on the mutual information distribution and then the algorithm described above would approach this curve instead of the actual distribution of a given data set. Current efforts are underway to generate this theoretical distribution.

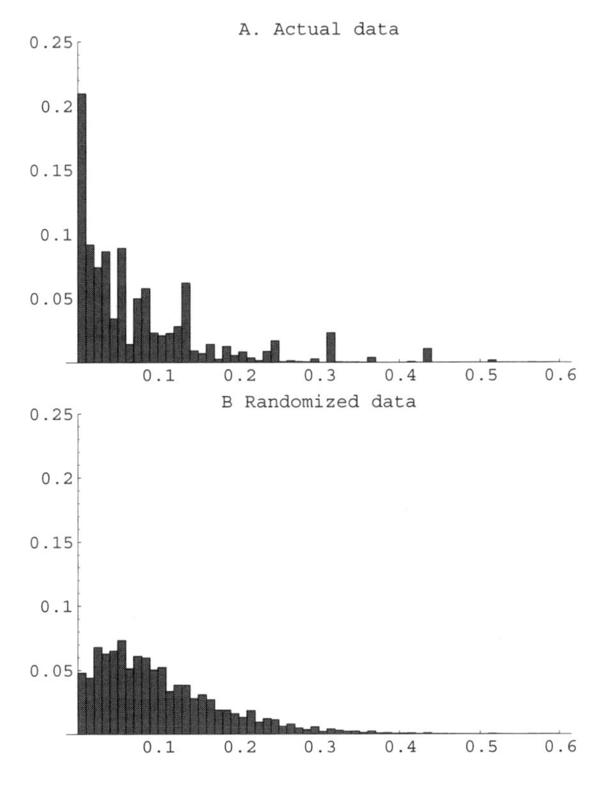

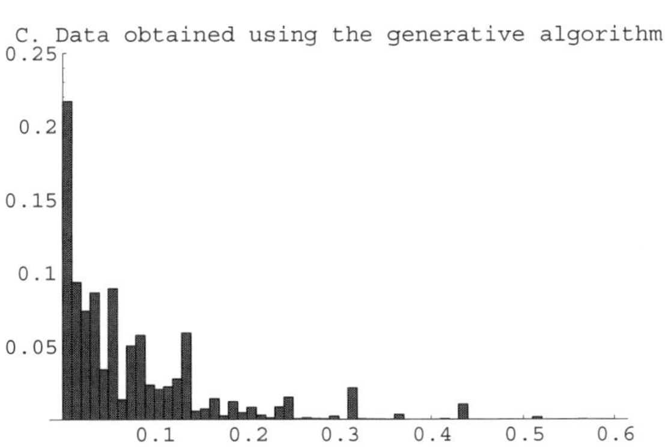

Figure 4.2 Upper panel: mutual information distribution from the Perou data set. Middle panel: the mutual information distribution obtained after simple randomization. Lower panel: the mutual information distribution obtained using a generative model.

References

Alon, U., Barkai, N., Notterman, D.A., Gish, K., Ybarra, S., Mack, D., and Levine, A.J. (1999). Broad patterns of gene expression revealed by clustering analysis of tumor and normal colon tissues probed by oligonucleotide arrays. *Proc. Natl. Acad. Sci. USA* 96(12):6745–50.

Butte, A.J. and Kohane, I.S. (2000). Mutual Information Relevance Networks: Functional Genomic Clustering Using Pairwise Entropy Measurements. *Pac. Symp. Biocomp. 2000* 5:415-426.

Chen, Y., Dougherty, E.R., and Bittner, M.L. (1997). Ratio-based decisions and the quantitative analysis of cDNA microarray images. *J. Biomed. Optics* 2:364–374.

Galitski, T., Saldanha, A.J., Styles, C.A., Lander, E.S., and Fink, G.R. (1999). Ploidy regulation of gene expression. *Science* 285:251–254.

Golub T.R., Slonim D.K., Tamayo P., Huard C., Gaasenbeek M., Mesirov J.P., Coller H., Loh M.L., Downing J.R., Caligiuri M.A., Bloomfield C.D., and Lander E.S. (1999). Molecular classification of cancer: class discovery and class prediction by gene expression monitoring. *Science* 286(5439):531–537.

Kauffman, S. (1993). *The origins of order*. Oxford University Press.

Klus, G., Song, A., Schick, A., Wahde, M., and Szallasi, Z. (2001). Mutual information analysis as a tool to assess the role of aneuploidy in the generation of cancer-associated differential gene expression patterns. *Pac. Symp. Biocomp. 2000* 6:42–51.

Khan J., Simon R., Bittner M., Chen Y., Leighton SB., Pohida T., Smith P.D., Jiang Y., Gooden G.C., Trent J.M., and Meltzer P. (1998). Gene expression profiling of alveolar rhabdomyosarcoma with cDNA microarrays. *Cancer Res.* 58(22):5009–5013.

Mitchell, M. (1995). *An introduction to genetic algorithms.* MIT Press.

Perou C.M., Jeffrey S.S., van de Rijn M., Rees C.A., Eisen M.B., Ross D.T., Pergamenschikov A., Williams C.F., Zhu S.X., Lee J.C., Lashkari D.,

Shalon D., Brown P.O., and Botstein D. (1999). Distinctive gene expression patterns in human mammary epithelial cells and breast cancers. *Proc. Natl. Acad. Sci. USA* 96(16):9212–7.

Szallasi, Z. (1998). Gene expression patterns and cancer. *Nature Biotech* 16:1292–1293.

Szallasi, Z., and Liang, S. (1998). Modeling the normal and neoplastic cell cycle with "realistic Boolean genetic networks": Their application for understanding carcinogenesis and assessing therapeutic strategies. *Pac. Symp. Biocomp. '98* 3:66–76.

Wahde, M., Klus, G.T., Bittner, M.L., Chen Y., and Szallasi, Z. (2001). Assessing the significance of consistently mis–regulated genes in cancer associated gene expression matrices. (manuscript in preparation)

5 Automated Reverse Engineering of Metabolic Pathways from Observed Data by Means of Genetic Programming

John R. Koza, William Mydlowec, Guido Lanza, Jessen Yu, and Martin A. Keane

Recent work has demonstrated that genetic programming is capable of automatically creating complex networks (e.g., analog electrical circuits, controllers) whose behavior is modeled by linear and non-linear continuous-time differential equations and whose behavior matches prespecified output values. The concentrations of substances participating in networks of chemical reactions are modeled by non-linear continuous-time differential equations. This chapter demonstrates that it is possible to automatically create (reverse engineer) a network of chemical reactions from observed time-domain data. Genetic programming starts with observed time-domain concentrations of substances and automatically creates both the topology of the network of chemical reactions and the rates of each reaction of a network such that the behavior of the automatically created network matches the observed time-domain data. Specifically, genetic programming automatically created a metabolic pathway involving four chemical reactions that consume glycerol and fatty acid as input, use ATP as a cofactor, and produce diacyl-glycerol as the final product. The metabolic pathway was created from 270 data points. The automatically created metabolic pathway contains three key topological features, including an internal feedback loop, a bifurcation point where one substance is distributed to two different reactions, and an accumulation point where one substance is accumulated from two sources. The topology and sizing of the entire metabolic pathway was automatically created using only the time-domain concentration values of diacyl-glycerol (the final product).

INTRODUCTION

A living cell can be viewed as a dynamical system in which a large number of different substances react continuously and non-linearly with one another. In order to understand the behavior of a continuous non-linear dynamical system with numerous interacting parts, it is usually insufficient to study behavior of each part in isolation. Instead, the behavior must usually be analyzed as a whole (Tomita et al., 1999).

Considerable amounts of time-domain data are now becoming available concerning the concentration of biologically important chemicals in living organisms. Such data include both gene expression data (obtained from microarrays) and data on the concentration of substances participating in metabolic pathways (Ptashne, 1992; McAdams and Shapiro, 1995; Loomis and Sternberg, 1995; Arkin et al., 1997; Yuh et al., 1998; Liang et al., 1998; Mendes and Kell, 1998; D'haeseleer et al., 1999).

The concentrations of substrates, products, and catalysts (e.g., enzymes) participating in chemical reactions are modeled by non-linear continuous-time differential equations, such as the Michaelis-Menten equations (Voit, 2000).

The question arises as to whether it is possible to start with observed time-domain concentrations of substances and automatically create both the topology of the network of chemical reactions and the rates of each reaction that produced the observed data — that is, to automatically reverse engineer the network from the data.

Genetic programming (Koza et al., 1999a) is a method for automatically creating a computer program whose behavior satisfies certain high-level requirements. Recent work has demonstrated that genetic programming can automatically create complex networks that exhibit prespecified behavior in areas where the network's behavior is governed by differential equations (both linear and non-linear).

For example, genetic programming is capable of automatically creating both the topology and sizing (component values) for analog electrical circuits (e.g., filters, amplifiers, computational circuits) composed of transistors, capacitors, resistors, and other components merely by specifying the circuit's output — that is, the output data values that would be observed if one already had the circuit. This reverse engineering of circuits from data is performed by genetic programming even though there is no general mathematical method for creating the topology and sizing of analog electrical circuits from the circuit's desired (or observed) behavior (Koza et al., 1999b). Seven of the automatically created circuits infringe on previously issued patents. Others duplicate the functionality of previously patented inventions in a novel way.

As another example, genetic programming is capable of automatically creating both the topology and sizing (tuning) for controllers composed of time-domain blocks such as integrators, differentiators, multipliers, adders, delays, leads, and lags merely by specifying the controller's effect on the to-be-controlled plant (Koza et al., 1999c, 2000a). This reverse engineering of controllers from data is performed by genetic programming even though there is no general mathematical method for creating the topology and sizing for controllers from the controller's behavior. Two of the automatically created controllers infringe on previously issued patents.

As yet another example, it is possible to automatically create antennas composed of a network of wires merely by specifying the antenna's high-level specifications (Comisky, Yu, and Koza 2000).

Our approach to the problem of automatically creating both the topology and sizing of a network of chemical reactions involves

(1) establishing a representation involving program trees (composed of functions and terminals) for chemical networks,

(2) converting each individual program tree in the population into an electrical circuit representing a network of chemical reactions,

(3) obtaining the behavior of the network of chemical reactions by simulating the electrical circuit,

(4) defining a fitness measure that measures how well the behavior of an individual network in the population matches the observed data, and

(5) applying genetic programming to breed a population of improving program trees using the fitness measure.

The implementation of our approach entails working with five different representations for a network of chemical reactions, namely

Reaction Network: Biochemists often use this representation (shown in Figure 5.1) to represent a network of chemical reactions. In this representation, the blocks represent chemical reactions and the directed lines represent flows of substances between reactions.

Program Tree: A network of chemical reactions can also be represented as a program tree whose internal points are functions and external points are terminals. This representation enables genetic programming to breed a population of programs in a search for a network of chemical reactions whose time-domain behavior concerning concentrations of final product substance(s) closely matches observed data.

Symbolic Expression: A network of chemical reactions can also be represented as a symbolic expression (S-expression) in the style of the LISP programming language. This representation is used internally by the run of genetic programming.

System of Non-Linear Differential Equations: A network of chemical reactions can also be represented as a system of non-linear differential equations.

Analog Electrical Circuit: A network of chemical reactions can also be represented as an analog electrical circuit (as shown in Figure 5.3). Representation of a network of chemical reactions as a circuit facilitates simulation of the network's time-domain behavior.

The goal is to automatically create (reverse engineer) *both* the topology and sizing of a network of chemical reactions.

The *topology* of a network of chemical reactions comprises (1) the number of substrates consumed by each reaction, (2) the number of products produced by each reaction, (3) the pathways supplying the substrates (either from external sources or other reactions) to the reactions, and (4) the pathways dispersing the reaction's products (either to other reactions or external outputs). The *sizing* of a network of chemical reactions consists of the numerical values representing the rates of each reaction.

We chose, as an illustrative problem, a network that incorporates three key topological features. These features include an internal feedback loop, a bifurcation point (where one substance is distributed to two different reactions), and an accumulation point (where one substance is accumulated from two sources). The particular chosen network is part of a phospholipid cycle, as presented in the E-CELL cell simulation model (Tomita et al., 1999). The network's external inputs are glycerol and fatty acid. The network's final product is diacyl-glycerol. The network's four reactions are catalyzed by Glycerol kinase (EC2.7.1.30), Glycerol-1-phosphatase (EC3.1.3.21), Acylglycerol lipase (EC3.1.1.23), and Triacylglycerol lipase (EC3.1.1.3). Figure **??** shows this network of chemical reactions with the correct rates of each reaction in parenthesis. The rates that are outside the parenthesis are the rates of the best individual from generation 225 of the run of genetic programming.

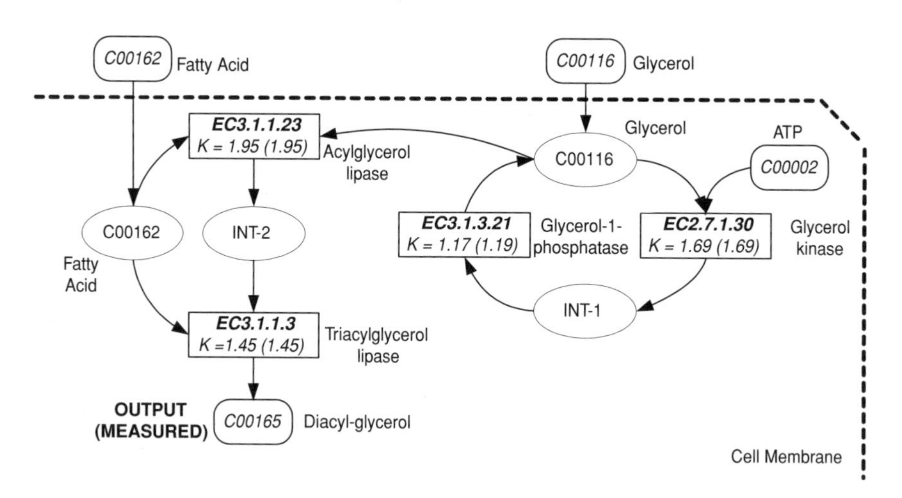

Figure 5.1 Network of chemical reactions involved in the phospholipid cycle

BACKGROUND ON GENETIC PROGRAMMING

Genetic programming (Koza, 1992, 1994a,b; Koza et al., 1999a,b; Koza and Rice, 1992) is a method for automatically creating a computer program whose behavior satisfies user-specified high-level requirements. Genetic programming is an extension of the genetic algorithm (Holland, 1992) in which the population being bred consists of computer programs. Genetic programming starts with a primordial ooze of thousands of randomly created computer programs (program trees) and uses the Darwinian principle of natural selection, crossover (sexual recombination), mutation, gene duplication, gene deletion, and certain mechanisms of developmental biology to breed a population of programs over a series of generations. Although there are many mathematical algorithms that solve problems by producing a set of numerical values, a run of genetic programming can create both a graphical structure and a set of numerical values. That is, genetic programming will produce not just numerical values, but the structure in which those numerical values reside.

Genetic programming breeds computer programs to solve problems by executing the following three steps:

(1) Generate an initial population of compositions (typically random) of the functions and terminals of the problem.

(2) Iteratively perform the following substeps (referred to herein as a generation) on the population of programs until the termination criterion has been satisfied:

(A)Execute each program in the population and assign it a fitness value using the fitness measure.

(B)Create a new population of programs by applying the following operations. The operations are applied to program(s) selected from the population with a probability based on fitness (with reselection allowed).

(i) Reproduction: Copy the selected program to the new population.

(ii) Crossover: Create a new offspring program for the new population by recombining randomly chosen parts of two selected programs.

(iii) Mutation: Create one new offspring program for the new population by randomly mutating a randomly chosen part of the selected program.

(iv) Architecture-altering operations: Select an architecture-altering operation from the available repertoire of such operations and create one new offspring program for the new population by applying the selected architecture-altering operation to the selected program.

(3) Designate the individual program that is identified by result designation (e.g., the best-so-far individual) as the result of the run of genetic programming. This result may be a solution (or an approximate solution) to the problem.

The individual programs that are evolved by genetic programming are typically multi-branch programs consisting of one or more result-producing branches and zero, one, or more automatically defined functions (subroutines).

The *architecture* of such a multi-branch program involves

(1) the total number of automatically defined functions,

(2) the number of arguments (if any) possessed by each automatically defined function, and

(3) if there is more than one automatically defined function in a program, the nature of the hierarchical references (including recursive references), if any, allowed among the automatically defined functions.

Architecture-altering operations enable genetic programming to automatically determine the number of automatically defined functions, the number of arguments that each possesses, and the nature of the hierarchical references, if any, among such automatically defined functions.

Additional information on genetic programming can be found in books such as (Banzhaf et al., 1998); books in the series on genetic programming from Kluwer Academic Publishers such as (Langdon, 1998); in edited collections of papers such as the *Advances in Genetic Programming* series of books from the MIT Press (Spector et al., 1999); in the proceedings of the Genetic Programming Conference (Koza et al., 1998); in the proceedings of the annual Genetic and Evolutionary Computation Conference (combining the annual Genetic Programming Conference and the International Conference on Genetic Algorithms) held starting in 1999 (Whitley et al., 2000); in the proceedings of the annual Euro-GP conferences held starting in 1998 (Poli et al., 2000); at web sites such as www.genetic-programming.org; and in the Genetic Programming and Evolvable Machines journal (from Kluwer Academic Publishers).

REPRESENTATION OF CHEMICAL REACTION NETWORKS

This section describes a method for representing a network of chemical reactions as a program tree suitable for use in a run of genetic programming. Each program tree represents an interconnected network of chemical reactions involving various substances. A chemical reaction may consume one or two substances and produce one or two substances. The consumed substances may be external input substances or intermediate substances produced by reactions. The chemical reactions, enzymes, and substances of a network may be represented by a program tree that contains

• internal nodes representing chemical reaction functions,
• internal nodes representing selector functions that select the reaction's first versus the reaction's second (if any) product,
• external points (leaves) representing substances that are consumed and produced by a reaction,
• external points representing enzymes that catalyze a reaction, and
• external points representing numerical constants (reaction rates).

Each program tree in the population is a composition of functions from the problem's function set and terminals from the problem's terminal set.

Repertoire of Functions

There are four chemical reaction functions and two selector functions.

The first argument of each chemical reaction (CR) function identifies the enzyme that catalyzes the reaction. The second argument specifies the reaction's rate. In addition, there are two, three, or four arguments specifying the substrate(s) and product(s) of the reaction. Table 5.1 shows the number of substrate(s) and product(s) and overall arity for each of the four chemical reaction functions. The runs in this chapter use a first-order and second-order rate law.

Table 5.1 Four chemical reaction functions

Function	Substrates	Products	Arity
CR_1_1	1	1	4
CR_1_2	1	2	5
CR_2_1	2	1	5
CR_2_2	2	2	6

Each function returns a list composed of the reaction's one or two products. The one-argument FIRST function returns the first of the one or two products produced by the function designated by its argument. The one-argument SECOND function returns the second of the two products (or, the first product, if the reaction produces only one product).

Repertoire of Terminals

Some terminals represent substances (input substances, intermediate substances created by reactions, or output substances). Other terminals represent the enzymes that catalyze the chemical reactions. Still other terminals represent numerical constants for the rate of the reactions.

Constrained Syntactic Structure

The trees are constructed in accordance with a constrained syntactic structure. The root of every result-producing branch must be a chemical reaction function. The enzyme that catalyzes a reaction always appears as the first argument of its chemical reaction function. A numerical value representing a reaction's rate always appears as the second argument of its chemical reaction function. The one or two input arguments to a chemical reaction function can be either a substance terminal or selector function (FIRST or SECOND). The result of having a selector function as an input argument is to create a cascade of reactions. The one or two output arguments to a chemical reaction function must be substance terminals. The argument to a one-argument selector function (FIRST or SECOND) is always a chemical reaction function.

Example

The chemical reactions, enzymes, and substances of a network of chemical reactions may be completely represented by a program tree that contains

- internal nodes representing chemical reaction functions,
- internal nodes representing selector functions that select the reaction's first versus the reaction's second (if any) product,
- external points (leaves) representing substances that are consumed and produced by a reaction,
- external points representing enzymes that catalyze a reaction, and
- external points representing numerical constants (reaction rates).

Each program tree in the population is a composition of functions from the following function set and terminals from the following terminal set.

Figure 5.2 shows a program tree that corresponds to the metabolic pathway of Figure 5.1. The program tree is presented in the style of the LISP programming language. The program tree (Figure 5.2) has two result-producing branches, RPB0 and RPB1. These two branches are connected by means of a connective PROGN function.

As can be seen, there are four chemical reaction functions in Figure 5.2. The first argument of each chemical reaction function is constrained to be an enzyme and the second argument is constrained to be a numerical rate. The remaining arguments are substances, such as externally supplied input substances, intermediate substances produced by reactions within the network, and the final output substance produced by the network. The remaining arguments of each chemical reaction function are marked, purely as a visual aid to the reader, by an arrow. An upward arrow indicates that the substance at the tail of the arrow points to a substrate of the reaction. An downward arrow indicates that the head of the arrow points to a product of the reaction.

There is a two-substrate, one-product chemical reaction function CR_2_1 in the lower left part of Figure 5.2. For this reaction, the enzyme is Acylglycerol lipase (EC3.1.1.23) (the first argument of this chemical reaction function); its rate is 1.95 (the second argument); its two substrates are fatty acid (C00162) (the third argument) and Glycerol (C00116) (the fourth argument); and its product is Monoacyl-glycerol (C01885) (the fifth argument).

There is a FIRST-PRODUCT function between the two chemical reaction functions in the left half of Figure 5.2. The FIRST-PRODUCT function selects the first of the two products of the lower CR_2_1 function. The line in the program tree from the lower chemical reaction function to the FIRST-PRODUCT function and the line between the FIRST-PRODUCT function and the higher CR_2_1 reaction means that when this tree is converted into a network of chemical reactions, the first (and, in this case, only) substance produced by the lower CR_2_1 reaction is a substrate to the higher reaction. In particular, the product of the lower reaction function (i.e., an intermediate substance called Monoacyl-glycerol) is the second of the two substrates to the higher chemical reaction function (i.e., the fourth argument of the higher function). Thus, although there is no return value for any branch or for the program tree as a whole, the return value(s) of all but the top chemical reaction function of a particular branch (as well the return values of a FIRST-PRODUCT function and a SECOND-PRODUCT function) define the flow of substances in the network of chemical reactions represented by the program tree.

Notice that the fatty acid (C00162) substance terminal appears as a substrate argument to both of these chemical reaction functions (in the left half of Figure 5.2 and also in the left half of Figure 5.1). The repetition of a substance terminal as a substrate argument in a program tree means that when the tree is converted into a network of chemical reactions, the available concentration of this particular substrate is distributed to two reactions in the network. That is, the repetition of a substance terminal as a substrate argument in a program tree corresponds to a bifurcation point where one substance is distributed to two different reactions in the network of chemical reactions represented by the program tree. There is another bifurcation point in this network of chemical reactions where Glycerol (C00116) appears as a substrate argument to both the two-substrate, one-product chemical reaction function CR_2_1 (in the lower left of Figure 5.2 and in the upper left part of Figure 5.1) and the two-substrate, two-product chemical reaction function CR_2_2 (in the upper right part of Figure 5.2 and in the upper right part of Figure 5.1).

Glycerol (C00116) has two sources in this network of chemical reactions. First, it is externally supplied (shown at the top right of Figure 5.1). Second, this substance is the product of the one-substrate, two-product chemical reaction function CR_1_2 (in the middle of Figure 5.1 and in the lower right of Figure 5.2). When a substance in a network has two or more sources (by virtue either of being externally supplied, by virtue of being a product of a reaction of a network, or any combination thereof), the substance is accumulated. When the program tree is converted into a network, all the sources of this substance are pooled. That is, there is an accumulation point for the substance.

Also, Glycerol (C00116) appears as part of an internal feedback loop consisting of two reactions, namely

- the one-substrate, two-product chemical reaction function CR_1_2 catalyzed by EC3.1.3.21 (in the middle of Figure 5.1 and in the lower right of Figure 5.2) and
- the two-substrate, two-product chemical reaction function CR_2_2 catalyzed by EC2.7.1.30 (in the upper right part of Figure 5.2 and in the right part of Figure 5.1).

The presence of an internal feedback loop is established in this network because of the following two features of this program tree:

- There exists a substance, namely sn-Glycerol-3-Phosphate (C00093) such that this substance
  - –is a product (sixth argument) that is produced by the two-substrate, two-product chemical reaction function CR_2_2 (catalyzed by EC2.7.1.30) in the upper right part of Figure 5.2, and
  - –is also a substrate that is consumed by the one-substrate, two-product chemical reaction function CR_1_2 (catalyzed by EC3.1.3.21) in the lower right part of Figure 5.2 that lies beneath the CR_2_2 function.
- There exists a second substance, namely glycerol (C00116), that
  - –is a product that is produced by the chemical reaction function CR_1_2 (catalyzed by EC3.1.3.21) and
  - –is a substrate that is consumed by the chemical reaction function CR_2_2 (catalyzed by EC2.7.1.30).

In summary, the network of Figure 5.2 contains the following three noteworthy topological features:

- an internal feedback loop in which Glycerol (C00116) is both consumed and produced in the loop,
- two bifurcation points (one where Glycerol is distributed to two different reactions and one where and fatty acid is distributed to two different reactions), and
- an accumulation point where one substance, namely Glycerol, is accumulated from two sources.

A Stanford University technical report provides additional details and explanatory figures (Koza, Mydlowec, Lanza, Yu, and Keane 2000).

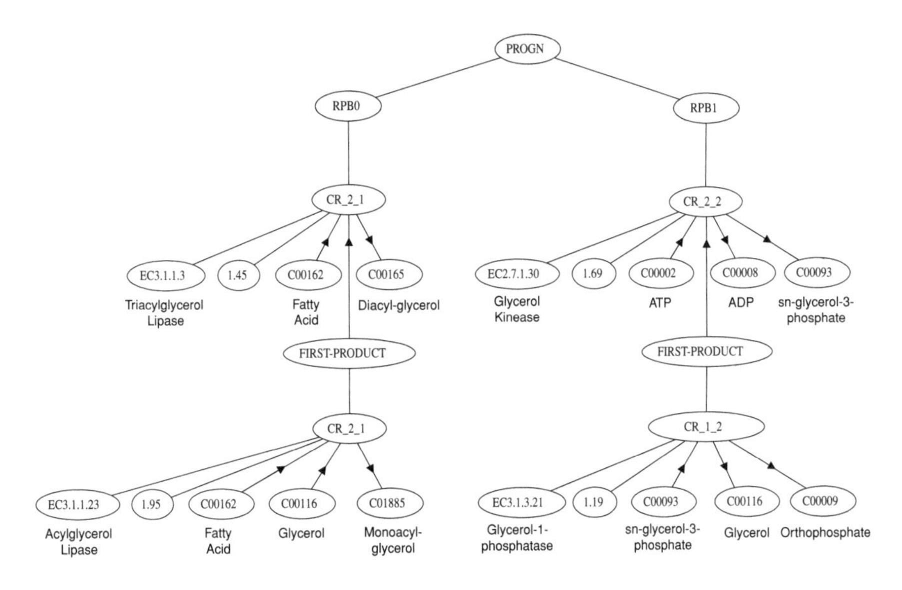

Figure 5.2 Program tree corresponding to metabolic pathway of Figure 5.1.

Figure 5.3 shows the electrical circuit corresponding to the network of Figure 5.1. The triangles in the figure represent integrators.

PREPARATORY STEPS

Six major preparatory steps are required before applying genetic programming: (1) determine the architecture of the program trees, (2) identify the functions, (3) identify the terminals, (4) define the fitness measure, (5) choose control parameters for the run, and (6) choose the termination criterion and method of result designation. For additional details, see (Koza et al., 2000b).

Program Architecture

Each program tree in the initial random population (generation 0) has one result-producing branch. In subsequent generations, the architecture-altering operations (patterned after gene duplication and gene deletion in nature) may insert and delete result-producing branches to particular individual program trees in the population. Each program tree may have four result-producing branches.

Function Set

The function set, F, consists of six functions.

F = {CR1_1, CR1_2, CR2_1, CR2_2, FIRST, SECOND}.

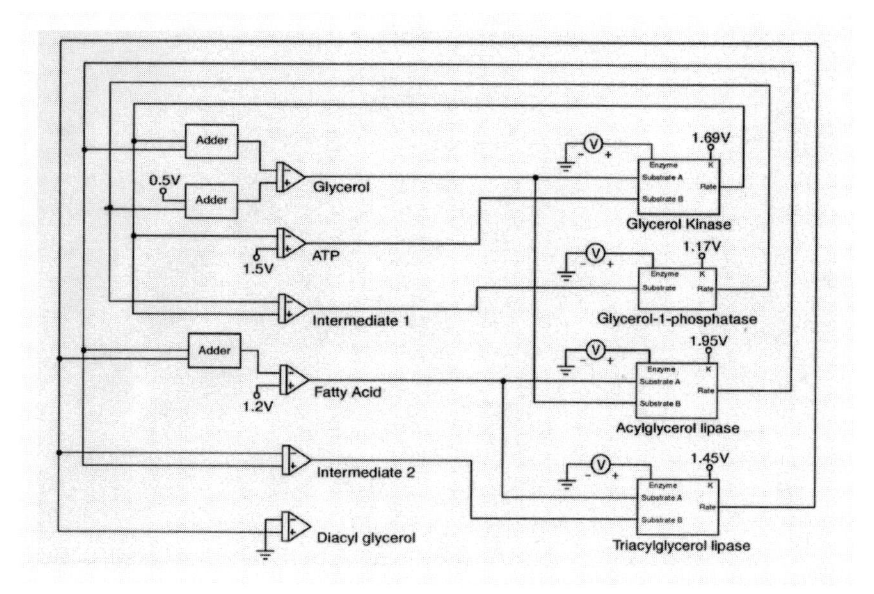

Figure 5.3 Electrical circuit corresponding to the chemical reaction network of Figure 5.1.

Terminal Set

The terminal set, T, is

$$T = \{\Re, C00116, C00162, C00002, C00165, INT\_1, INT\_2, INT\_3, EC2\_7\_1\_30, EC3\_1\_3\_21, EC3\_1\_1\_23, EC3\_1\_1\_3\}.$$

$\Re$ denotes a perturbable numerical value. In the initial random generation (generation 0) of a run, each perturbable numerical value is set, individually and separately, to a random value in a chosen range (from 0.0 and 2.0 here).

In the illustrative problem herein, C00116 is the concentration of glycerol. C00162 is the concentration of fatty acid. These two substances are inputs to the illustrative overall network of interest herein. C00002 is the concentration of the cofactor ATP. C00165 is the concentration of diacylglycerol. This substance is the final product of the illustrative network herein. INT_1, INT_2, and INT_3 are the concentrations of intermediate substances 1, 2, and 3 (respectively).

INT_1, INT_2, and INT_3 are the concentrations of intermediate substances 1, 2, and 3 (respectively).

EC2_7_1_30, EC3_1_3_21, EC3_1_1_23, and EC3_1_1_3 are enzymes.

Fitness Measure

Genetic programming is a probabilistic algorithm that searches the space of compositions of the available functions and terminals under the guidance of a fitness measure. In order to evaluate the fitness of an individual program tree in the population, the program tree is converted into a directed graph representing the network. The result-producing branches are executed from left to right. The functions in a particular result-producing branch are executed in a depth-first manner. One reactor (representing the concentration of the substances participating in the reaction) is inserted into the network for each chemical reaction function that is encountered in a branch. The reactor is labeled with the reaction's enzyme and rate. A directed line entering the reactor is added for each of the reaction's one or two substrate(s). A directed line leaving the reactor is added for each of the reaction's one or two product(s). The first product of a reaction is selected whenever a FIRST function is encountered in a branch. The second product of a reaction is selected whenever a SECOND function is encountered in a branch.

After the network is constructed, the pathway is converted into an electrical circuit. A SPICE netlist is then constructed to represent the electrical circuit. We provide SPICE with subcircuit definitions to implement all the chemical reaction equations. This SPICE netlist is wrapped inside an appropriate set of SPICE commands to carry out analysis in the time domain (described below). The electrical circuit is then simulated using our modified version of the original 217,000-line SPICE3 simulator (Quarles et al., 1994). We have embedded our modified version of SPICE as a submodule within our genetic programming system.

Each individual chemical reaction network is exposed to nine time-domain signals (table 2) representing the time-varying concentrations of four enzymes (EC2.7.1.30, EC3.1.3.21, EC3.1.1.23, and EC3.1.1.3) over 30 half-second time steps. None of these time series patterns are extreme. Each has been structured so as to vary the concentrations between 0 and 2.0 in a pattern to which a living cell might conceivably be exposed.

There are a total of 270 data points. The data was obtained from the E-CELL cell simulation model (Tomita et al., 1999; Voit, 2000).

The concentrations of all intermediate substances and the network's final product are 0 at time step 0.

For the runs in this paper, Glycerol (C00116), Fatty acid (C00162), and ATP (C00002) are externally supplied at a constant rate (table 3). That is, these values are not subject to evolutionary change during the run.

Fitness is the sum, over the 270 fitness cases, of the absolute value of the difference between the concentration of the end product of the individual reaction network and the observed concentration of diacylglycerol (C00165). The smaller the fitness, the better. An individual that cannot be simulated by SPICE is assigned a high penalty value of fitness

Table 5.2 Variations in the levels of the four enzymes

Signal	EC2.7.1.30	EC3.1.3.21	EC.1.1.23	EC3.1.1.3
1	Slope-Up	Sawtooth	Step-Down	Step-Up
2	Slope-Down	Step-Up	Sawtooth	Step-Down
3	Step-Down	Slope-Up	Slope-Down	Step-Up
4	Step-Up	Slope-Down	Step-Up	Step-Down
5	Sawtooth	Step-Down	Slope-Up	Step-Up
6	Sawtooth	Step-Down	Knock-Out	Step-Up
7	Sawtooth	Knock-Out	Slope-Up	Step-Down
8	Knock-Out	Step-Down	Slope-Up	Sawtooth
9	Step-Down	Slope-Up	Sawtooth	Knock-Out

Table 5.3 Rates for three externally supplied substances

Substance	Rate
Glycerol (C00116)	0.5
Fatty acid (C00162)	1.2
ATP (C00002)	1.5

(10^8). The number of hits is defined as the number of fitness cases (0 to 270) for which the concentration of the measured substances is within 5% of the observed data value.

See (Koza et al., 2000b) for additional details.

Control Parameters for the Run

The population size, M, is 100,000. A generous maximum size of 500 points (for functions and terminals) was established for each result-producing branch. The percentages of the genetic operations for each generation is 58.5% one-offspring crossover on internal points of the program tree other than perturbable numerical values, 6.5% one-offspring crossover on points of the program tree other than perturbable numerical values, 1% mutation on points of the program tree other than perturbable numerical values, 20% mutation on perturbable numerical values, 10% reproduction, 3% branch creation, and 2% subroutine deletion. The other parameters are the default values that we apply to a broad range of problems (Koza et al., 1999a).

Termination

The run was manually monitored and manually terminated when the fitness of many successive best-of-generation individuals appeared to have reached a plateau.

Implementation on Parallel Computing System

We used a home-built Beowulf-style (Sterling et al., 1999; Koza et al., 1999a) parallel cluster computer system consisting of 1,000 350 MHz Pentium II processors (each accompanied by 64 megabytes of RAM). The system has a 350 MHz Pentium II computer as host. The processing nodes are connected with a 100 megabit-per-second Ethernet. The processing nodes and the host use the Linux operating system. The distributed genetic algorithm with unsynchronized generations and semi-isolated subpopulations was used with a subpopulation size of $Q = 500$ at each of $D = 1{,}000$ demes. As each processor (asynchronously) completes a generation, four boatloads of emigrants from each subpopulation are dispatched to each of the four toroidally adjacent processors. The 1,000 processors are hierarchically organized. There are $5 \times 5 = 25$ high-level groups (each containing 40 processors). If the adjacent node belongs to a different group, the migration rate is 2% and emigrants are selected based on fitness. If the adjacent node belongs to the same group, emigrants are selected randomly and the migration rate is 5% (10% if the adjacent node is in the same physical box).

RESULTS

The population for the initial random generation (generation 0) of a run of genetic programming is created at random. The fitness of the best individual (Figure 5.4) from generation 0 is 86.4. This individual scores 126 hits (out of 270). Substance C00162 (fatty acid) is used as an input substance to this metabolic pathway; however, glycerol (C00116) and ATP (C00002) are not. Two of the four available reactions (EC 3.1.1.23 and EC 3.1.1.3) are used. However; a third reaction (EC 3.1.3.21) consumes a non-existent intermediate substance (INT_2) and the fourth reaction (EC 2.7.1.30) is not used at all. This metabolic pathway contains one important topological feature, namely the bifurcation of C00162 to two different reactions. However, this metabolic pathway does not contain any of the other important topological features of the correct metabolic pathway.

In generation 10, the fitness of the best individual (Figure 5.5) is 64.0. This individual scores 151 hits. This metabolic pathway is superior to the best individual of generation 0 in that it uses both C00162 (fatty acid) and glycerol (C00116) as external inputs. However, this metabolic pathway does not use ATP (C00002). This metabolic pathway is also defective in that it contains only two of the four reactions.

In generation 25, the fitness of the best individual (figure 5.6) is 14.3. This individual scores 224 hits. This metabolic pathway contains all four of the available reactions. This metabolic pathway is more complex than previous best-of-generation individuals in that it contains two topological features not previously seen. First, this metabolic pathway contains an internal feedback loop in which one substance (glycerol C00116) is con-

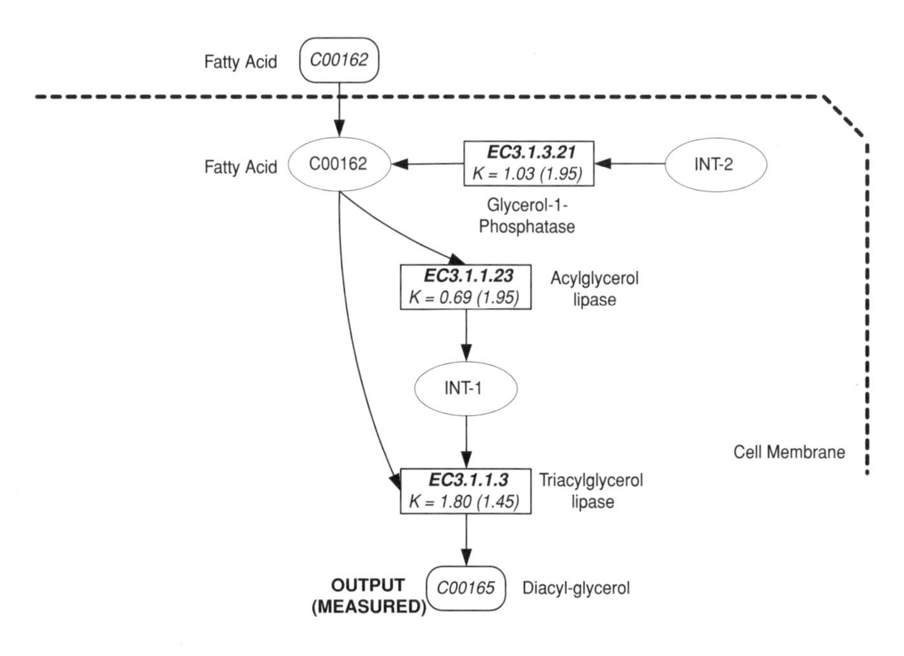

Figure 5.4 Best of generation 0

sumed by one reaction (catalyzed by enzyme EC 2.7.1.30), produced by another reaction (catalyzed by enzyme EC 3.1.3.21), and then supplied as a substrate to the first reaction. Second, this metabolic pathway contains a place where there is an addition of quantities of one substance. Specifically, glycerol (C00116) comes from the reaction catalyzed by enzyme EC 3.1.3.21 and is also externally supplied. This metabolic pathway also contains two substances (C00116 and C00162) where a substance is bifurcated to two different reactions.

$$\frac{d[ATP]}{dt} = 1.5 - 1.69[C00116][C00002][EC2.7.1.30] \tag{5.1}$$

In generation 120, the fitness of the best individual (Figure 5.7) is 2.33. The cofactor ATP (C00002) appears as an input to this metabolic pathway. This pathway has the same topology as the correct network. However, the numerical values (sizing) are not yet correct and this individual scores only 255 hits.

The best-of-run individual (Figure 5.1) appears in generation 225. Its fitness is almost zero (0.054). This individual scores 270 hits (out of 270). In addition to having the same topology as the correct metabolic pathway, the rate constants of three of the four reactions match the correct rates (to three significant digits) while the fourth rate differs by only about 2% from the correct rate (i.e., the rate of EC 3.1.3.21 is 1.17 compared with 1.19 for the correct network).

In the best-of-run network from generation 225, the rate of production

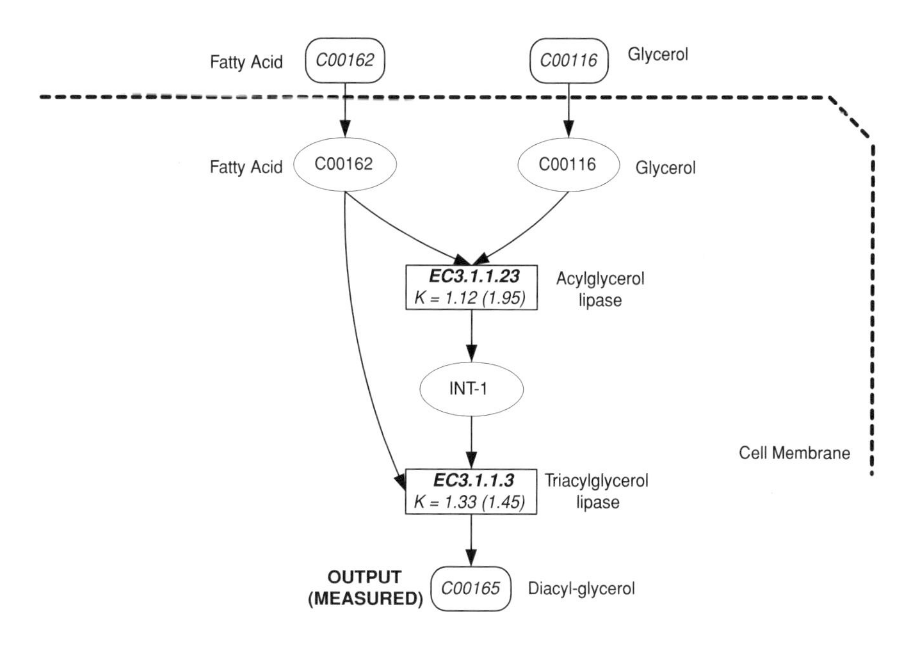

Figure 5.5 Best of generation 10

of the network's final product, diacyl-glycerol (C00165), is given by

$$\frac{d[C00165]}{dt} = 1.45[C00162][INT_2][EC3.1.1.3] \tag{5.2}$$

Note that genetic programming has correctly determined that the reaction that produces the network's final product diacyl-glycerol (C00165) has two substrates and one product; it has correctly identified enzyme EC3.1.1.3 as the catalyst for this final reaction; it has correctly determined the rate of this final reaction as 1.45; and it has correctly identified the externally supplied substance, fatty acid (C00162), as one of the two substrates for this final reaction. None of this information was supplied *a priori* to genetic programming.

Of course, genetic programming has no way of knowing that biochemists call the intermediate substance (INT_2) by the name Monoacylglycerol (C01885) (as indicated in Figure 5.1). It has, however, correctly determined that an intermediate substance is needed as one of the two substrates of the network's final reaction and that this intermediate substance should, in turn, be produced by a particular other reaction (described next).

In the best-of-run network from generation 225, the rate of production and consumption of the intermediate substance INT_2 is given by

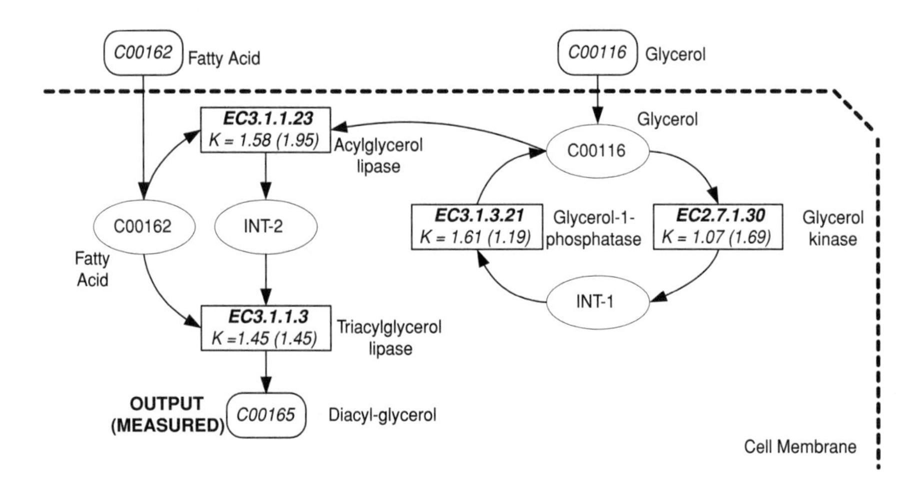

Figure 5.6 Best of generation 25

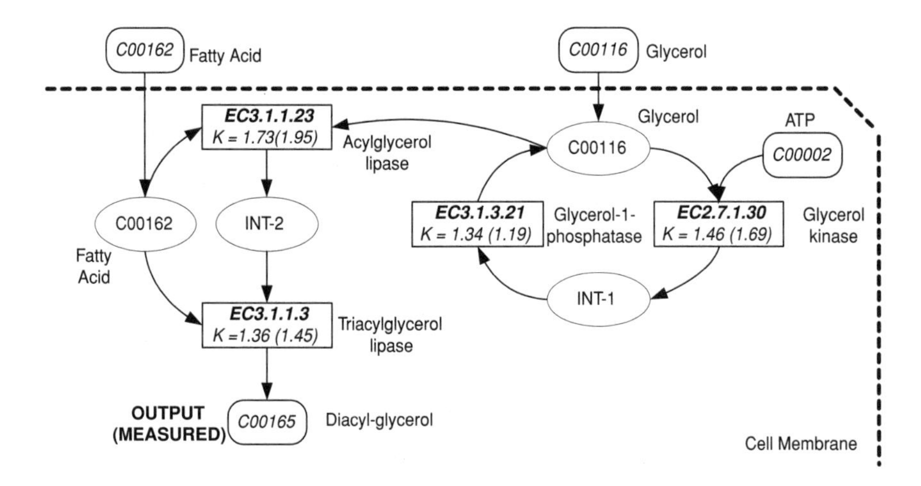

Figure 5.7 Best of generation 120

$$\frac{d[INT_2]}{dt} = 1.95[C00162][C00116][EC3.1.1.23] - 1.45[C00162][INT_2][EC3.1.1.3] \quad (5.3)$$

Again, genetic programming has correctly determined that the reaction that produces the intermediate substance (INT_2) has two substrates and one product; it has correctly identified enzyme EC3.1.1.23 as the catalyst for this reaction; it has correctly determined the rate of this reaction as 1.95; it has correctly identified two externally supplied substances, fatty acid (C00162) and glycerol (C00116), as the two substrates for this reaction.

In the best-of-run network from generation 225, the rate of production and consumption of the intermediate substance INT_1 in the internal feedback loop is given by

$$\frac{d[INT_1]}{dt} = 1.69[C00116][C00002][EC2.7.1.30] - 1.17[INT_1][EC3.1.3.21] \tag{5.4}$$

Note that the numerical rate constant of 1.17 in the above equation is slightly different from the correct rate (as shown in Figure 5.1).

Here again, genetic programming has correctly determined that the reaction that produces the intermediate substance (INT_1) has two substrates and one product; it has correctly identified enzyme EC2.7.1.30 as the catalyst for this reaction; it has almost correctly determined the rate of this reaction to be 1.17 (whereas the correct rate is 1.19, as shown in Figure 5.1); it has correctly identified two externally supplied substances, glycerol (C00116) and the cofactor ATP (C00002), as the two substrates for this reaction.

Genetic programming has no way of knowing that biochemists call the intermediate substance (INT_1) by the name sn-Glycerol-3-Phosphate (C00093) (as indicated in Figure 5.1). Genetic programming has, however, correctly determined that an intermediate substance is needed as the single substrate of the reaction catalyzed by Glycerol-1-phosphatase (EC3.1.3.21) and that this intermediate substance should, in turn, be produced by the reaction catalyzed by Glycerol kinase (EC2.7.1.30).

In the best-of-run network from generation 225, the rate of supply and consumption of ATP (C00002) is

$$\frac{d[ATP]}{dt} = 1.5 - 1.69[C00116][C00002][EC2.7.1.30] \tag{5.5}$$

The rate of supply and consumption of fatty acid (C00162) in the best-of-run network is

$$\begin{aligned}\frac{d[C00162]}{dt} = 1.2 - 1.95[C00162][C00116][EC3.1.1.23] \\ -1.45[C00162][INT_2][EC3.1.1.3]\end{aligned} \tag{5.6}$$

The rate of supply, consumption, and production of glycerol (C00116) in the best-of-run network is

$$\begin{aligned}\frac{d[C00116]}{dt} &= 0.5 + 1.17[INT_1][EC3.1.3.21] \\ &-1.69[C00116][C00002][EC2.7.1.30] \\ &-1.95[C00162][C00116][EC3.1.1.23]\end{aligned} \tag{5.7}$$

Again, note that the numerical rate constant of 1.17 in the above equation is slightly different from the correct rate (as shown in Figure 5.1).

Notice the internal feedback loop in which C00116 is both consumed and produced.

In summary, driven only by the time-domain concentration values of the final product C00165 (diacyl-glycerol), genetic programming created both the topology and sizing for an entire metabolic pathway whose time-domain behavior closely matches that of the naturally occurring pathway, including

- the total number of reactions in the network,
- the number of substrate(s) consumed by each reaction,
- the number of product(s) produced by each reaction,
- an indication of which enzyme (if any) acts as a catalyst for each reaction,
- the pathways supplying the substrate(s) (either from external sources or other reactions in the network) to each reaction,
- the pathways dispersing each reaction's product(s) (either to other reactions or external outputs),
- the number of intermediate substances in the network,
- emergent topological features such as
  - –internal feedback loops,
  - –bifurcation points,
  - –accumulation points, and
- numerical rates (sizing) for all reactions.

Genetic programming did this using only the 270 time-domain concentration values of the final product C00165 (diacyl-glycerol).

For additional details, see (Koza et al., 2000b).

CONCLUSION

Genetic programming automatically created a metabolic pathway involving four chemical reactions that took in glycerol and fatty acid as input, used ATP as a cofactor, and produced diacyl-glycerol as its final product. The metabolic pathway was created from 270 data points. The automatically created metabolic pathway contains three key topological features, including an internal feedback loop, a bifurcation point where one substance is distributed to two different reactions, and an accumulation point where one substance is accumulated from two sources. This example demonstrates the principle that it is possible to reverse engineer a metabolic pathway using only observed data for the concentration values of the pathway's final product.

FUTURE WORK

Numerous directions for future work are suggested by the work described herein.

Improved Program Tree Representation

Although the representation used herein yielded the desired results, the authors believe that alternative representations for the program tree (i.e., the function set, terminal set, and constrained syntactic structure) would significantly improve efficiency of the search. The authors are currently contemplating a developmental approach.

Minimum Amount of Data Needed

The work in this chapter has not addressed the important question of the minimal number of data points necessary to automatically create a correct metabolic pathway or the question whether the requisite amount of data is available in practical situations.

Opportunities to Use Knowledge

There are numerous opportunities to incorporate and exploit preexisting knowledge about chemistry and biology in the application of the methods described in this chapter.

The chemical reactions functions used in this chapter (i.e., CR_1_1, CR_1_2, CR_2_1, CR_2_2) are intentionally open-ended in the sense that they permit great flexibility and variety in the networks that can be created by the evolutionary process. However, there is a price, in terms of efficiency of the run, that is paid for this flexibility and generality. Alternative chemical reaction functions that advantageously incorporate preexisting knowledge might be defined and included in the function set.

For example, a particular substrate, a particular product, or both might be made part of the definition of a new chemical reaction function. For example, a variant of the CR_2_2 chemical reaction function might be defined in which ATP is hard-wired as one of the substrates and ADP is hard-wired as one of products. This new chemical reaction function would have only one free substrate argument and one free product argument. This new chemical reaction function might be included in the function set in addition to (and conceivably in lieu of) the more general and open-ended CR_2_2 chemical reaction function. This new chemical reaction function would exploit the well-known fact that there are a number of biologically important and biologically common reactions that employ ATP as one of its two substrates and produce ADP as one of its products.

Similarly, a particular enzyme might be made part of the definition of a new chemical reaction function. That is, a chemical reaction function with k substrates and j products might be defined in which a particular enzyme is hard-wired. This new chemical reaction function would not possess an argument for specifying the enzyme. This new chemical reaction function would exploit knowledge of the arity of reactions catalyzed by a particular enzyme.

Also, a known rate might be made part of the definition of a new chemical reaction function. This approach might be particularly useful in combination with other alternatives mentioned above.

Designing Alternative Metabolisms

Mittenthal et al. (1998) have presented a method for generating alternative biochemical pathways. They illustrated their method by generating diverse alternatives to the non-oxidative stage of the pentose phosphate pathway. They observed that the naturally occurring pathway is especially favorable in several respects to the alternatives that they generated. Specifically, the naturally occurring pathway has a comparatively small number of steps, does not use any reducing or oxidizing compounds, requires only one ATP in one direction of flux, and does not depend on recurrent inputs.

Mendes and Kell (1998) have also suggested that novel metabolic pathways might be artificially constructed.

It would appear that genetic programming could also be used to generate diverse alternatives to naturally occurring pathways. Conceivably, realizable alternative metabolisms might emerge from such evolutionary runs.

In one approach, the fitness measure in a run of genetic programming might be oriented toward duplicating the final output(s) of the naturally occurring pathway (as was done in this chapter). However, instead of harvesting only the individual from the population with the very best value of fitness, individuals that achieve a slightly poorer value of fitness could be examined to see if they simultaneously possess other desirable characteristics.

In a second approach, the fitness measure in a run of genetic programming might be specifically oriented to factors such as the pathway's efficiency or use or non-use of certain specified reactants or enzymes.

In a third approach, the fitness measure in a run of genetic programming might be specifically oriented toward achieving novelty. Genetic programming has previously been used as an invention machine by employing a two-part fitness measure that incorporates both the degree to which an individual in the population satisfies the certain performance requirements and the degree to which the individual does not possess the key characteristics of previously known solutions (Koza et al., 1999a,c).

ACKNOWLEDGEMENTS

Douglas B. Kell of the University of Wales made helpful comments on a draft of this material.

References

Arkin, A., Peidong, S., and Ross, J. (1997). A test case of correlation metric construction of a reaction pathway from measurements. *Science* 277:1275–1279.

Banzhaf, W., Nordin, P., Keller, R.E., and Francone, F.D. (1998). *Genetic Programming – An Introduction*. San Francisco, CA: Morgan Kaufmann and Heidelberg: dpunkt.

Comisky, W., and Yu, J., and Koza, J. (2000). Automatic synthesis of a wire antenna using genetic programming. *Late Breaking Papers at the 2000 Genetic and Evolutionary Computation Conference, Las Vegas, Nevada*.

D'haeseleer, P., Wen, X., Fuhrman, S., and Somogyi, R. (1999). Linear modeling of mRNA expression levels during CNS development and injury. *Proc. Paficic Symposium on Biocomputing'99* pp.41–52.

Holland, J.H. (1992) *Adaptation in Natural and Artificial Systems: An Introductory Analysis with Applications to Biology, Control, and Artificial Intelligence*. Ann Arbor, MI: University of Michigan Press 1975. Second edition. Cambridge, MA: The MIT Press 1992.

Koza, J.R. (1992). *Genetic Programming: On the Programming of Computers by Means of Natural Selection*. MIT Press.

Koza, J.R. (1994a). *Genetic Programming II: Automatic Discovery of Reusable Programs*. MIT Press.

Koza, J.R. (1994b). *Genetic Programming II Videotape: The Next Generation*. MIT Press.

Koza, J.R., Banzhaf, W., Chellapilla, K., Deb, K., Dorigo, M., Fogel, D.B., Garzon, M.H., Goldberg, D.E., Iba, H., and Riolo, R. (editors). (1998). *Genetic Programming 1998: Proceedings of the Third Annual Conference*. San Francisco, CA: Morgan Kaufmann.

Koza, J.R., Bennett III, F.H, Andre, D, and Keane, M.A. (1999a). *Genetic Programming III: Darwinian Invention and Problem Solving*. San Francisco, CA: Morgan Kaufmann.

Koza, J.R., Bennett III, F.H, Andre, D., Keane, M.A., and Brave, S. (1999b). *Genetic Programming III Videotape: Human-Competitive Machine Intelligence*. San Francisco, CA: Morgan Kaufmann.

Koza, J.R., Keane, M.A., Yu, J., Bennett III, F.H., Mydlowec, W., and Stiffelman, O. (1999c). Automatic synthesis of both the topology and parameters for a robust controller for a non-minimal phase plant and a three-lag plant by means of genetic programming. *Proceedings of 1999 IEEE Conference on Decision and Control* pp.5292–5300.

Koza, J.R., Keane, M.A., Yu, J., Bennett III, F.H., and Mydlowec, W. (2000a). Automatic creation of human-competitive programs and controllers by means of genetic programming. *Genetic Programming and Evolvable Machines* 1:121–164.

Koza, J.R., Mydlowec, W., Lanza, G., Yu, J., and Keane, M.A. (2000b). *Reverse Engineering and Automatic Synthesis of Metabolic Pathways from Observed Data Using Genetic Programming*. Stanford Medical Informatics Technical Report SMI-2000-0851.

Koza, J.R. and Rice, J.P. (1992). *Genetic Programming: The Movie*. Cambridge, MA: MIT Press.

Liang, S., Fuhrman, S., and Somogyi, R. (1998). REVEAL: A general reverse engineering algorithm for inference of genetic network architecture. *Proc. Pacific Symposium on Biocomputing '98* pp.18-29.

Langdon, W.B. (1998). *Genetic Programming and Data Structures: Genetic Programming + Data Structures = Automatic Programming!* Amsterdam: Kluwer.

Loomis, W.F. and Sternberg, P.W. (1995). Genetic networks. *Science* 269:649.

McAdams, H.H. and Shapiro, L. (1995). Circuit simulation of genetic networks. *Science* 269:650-656.

Mendes, P. and Kell, D.B. (1998). Non-linear optimization of biochemical pathways: Applications to metabolic engineering and parameter estimation. *Bioinformatics* 14(10):869–883.

Mittenthal, J.E., Ao, Y., Bertrand C., and Scheeline, A. (1998). Designing metabolism: Alternative connectivities for the pentose phosphate pathway. *Bulletin of Mathematical Biology* 60:815–856.

Poli, R., Banzhaf, W., Langdon, W.B., Miller, J., Nordin, P., and Fogarty, T.C. (2000). *Genetic Programming: European Conference, EuroGP 2000, Edinburgh, Scotland, UK, April 2000, Proceedings*. Lecture Notes in Computer Science. Volume 1802. Berlin, Germany: Springer-Verlag.

Ptashne, M. (1992). *A Genetic Switch: Phage λ and Higher Organisms.* Second Edition. Cambridge, MA: Cell Press and Blackwell Scientific Publications.

Quarles, T., Newton, A.R., Pederson, D.O., and Sangiovanni-Vincentelli, A. (1994). *SPICE 3 Version 3F5 User's Manual.* Department of Electrical Engineering and Computer Science, University of California. Berkeley, CA.

Spector, Lee, Langdon, William B., O'Reilly, Una-May, and Angeline, Peter (editors). (1999). *Advances in Genetic Programming 3.* Cambridge, MA: The MIT Press.

Sterling, T.L., Salmon, J., and Becker, D.J., and Savarese, D.F. (1999). *How to Build a Beowulf: A Guide to Implementation and Application of PC Clusters.* Cambridge, MA: MIT Press.

Tomita, M., Hashimoto, K., Takahashi, K., Shimizu, T.S., Matsuzaki, Y., Miyoshi, F., Saito, K., Tanida, S., Yugi, K., Venter, J.C., Hutchison, C.A. (1999). E-CELL: Software environment for whole cell simulation. *Bioinformatics* 15(1):72–84.

Voit, E.O. (2000). *Computational Analysis of Biochemical Systems.* Cambridge: Cambridge University Press.

Whitley, D., Goldberg, D., Cantu-Paz, E., Spector, L., Parmee, I., and Beyer, H.-G. (editors). (2000). *GECCO-2000: Proceedings of the Genetic and Evolutionary Computation Conference, July 10 - 12, 2000, Las Vegas, Nevada.* San Francisco: Morgan Kaufmann Publishers.

Yuh, C.-H., Bolouri, H., and Davidson, E.H. (1998). Genomic cis-regulatory logic: Experimental and computational analysis of a sea urchin gene. *Science* 279:1896–1902.

Part III

Software for Modeling and Simulation

6 The ERATO Systems Biology Workbench: An Integrated Environment for Multiscale and Multitheoretic Simulations in Systems Biology

Michael Hucka, Andrew Finney, Herbert Sauro,
Hamid Bolouri, John Doyle, and Hiroaki Kitano

Over the years, a variety of biochemical network modeling packages have been developed and used by researchers in biology. No single package currently answers all the needs of the biology community; nor is one likely to do so in the near future, because the range of tools needed is vast and new techniques are emerging too rapidly. It seems unavoidable that, for the foreseeable future, systems biology researchers are likely to continue using multiple packages to carry out their work.

In this chapter, we describe the ERATO *Systems Biology Workbench* (SBW) and the *Systems Biology Markup Language* (SBML), two related efforts directed at the problems of software package interoperability. The goal of the SBW project is to create an integrated, easy-to-use software environment that enables sharing of models and resources between simulation and analysis tools for systems biology. SBW uses a modular, plug-in architecture that permits easy introduction of new components. SBML is a proposed standard XML-based language for representing models communicated between software packages; it is used as the format of models communicated between components in SBW.

INTRODUCTION

The goal of the ERATO *Systems Biology Workbench* (SBW) project is to create an integrated, easy-to-use software environment that enables sharing of models and resources between simulation and analysis tools for systems biology. Our initial focus is on achieving interoperability between seven leading simulations tools: *BioSpice* (Arkin, 2001), *DBSolve* (Goryanin, 2001; Goryanin et al., 1999), *E-Cell* (Tomita et al., 1999, 2001), *Gepasi* (Mendes, 1997, 2001), *Jarnac* (Sauro, 1991; Sauro and Fell, 2000), *StochSim* (Bray et al., 2001; Morton-Firth and Bray, 1998), and *Virtual Cell* (Schaff et al., 2000, 2001). Our long-term goal is to develop a flexible and adaptable environronment that provides (1) the ability to interact seamlessly with a variety of software tools that implement different approaches to modeling, parameter analysis, and other related tasks, and (2) the ability to interact with biologically-oriented databases containing data, models and other relevant information.

In the sections that follow, we describe the Systems Biology Workbench project, including our motivations and approach, and we summarize our current design for the Workbench software environment. We also discuss the *Systems Biology Markup Language* (SBML), a model description language that serves as the common substrate for communications between components in the Workbench. We close by summarizing the current status of the project and our future plans.

Motivations for the Project

The staggering volume of data now emerging from molecular biotechnology leave little doubt that extensive computer-based modeling, simulation and analysis will be critical to understanding and interpreting the data (e.g., Abbott, 1999; Gilman, 2000; Popel and Winslow, 1998; Smaglik, 2000). This has lead to an explosion in the development of computer tools by research groups across the world. Example application areas include the following:

- Filtering and preparing data (e.g., gene expression micro- and macroarray image processing and clustering/outlier identification), as well as performing regression and pattern-extraction;
- Database support, including remote database access and local data storage and management (e.g., techniques for combining gene expression data with analysis of gene regulatory motifs);
- Model definition using graphical model capture and/or mathematical description languages, as well as model preprocessing and translation (e.g., capturing and describing the three-dimensional structure of subcellular structures, and their change over time);
- Model computation and analysis, including parameter optimization, bifurcation/sensitivity analysis, diffusion/transport/buffering in complex 3-D structures, mixed stochastic-deterministic systems, differential-algebraic systems, qualitative-qualitative inference, and so on; and
- Data visualization, with support for examining multidimensional data, large data sets, and interactive steering of ongoing simulations.

This explosive rate of progress in tool development is exciting, but the rapid growth of the field has been accompanied by problems and pressing needs. One problem is that simulation models and results often cannot be compared, shared or re-used directly because the tools developed by different groups often are not compatible with each other. As the field of systems biology matures, researchers increasingly need to communicate their results as computational models rather than box-and-arrow diagrams. But they also need to reuse each other's published and curated models as library elements in order to succeed with large-scale efforts (e.g., the Alliance for Cellular Signaling, Gilman, 2000; Smaglik, 2000). These needs require that models implemented in one software package be portable to other software packages, to maximize public understanding and to allow building up libraries of curated computational models.

A second problem is that software developers often end up duplicating each other's efforts when implementing different packages. The reason is that individual software tools typically are designed initially to address a specific set of issues, reflecting the expertise and preferences of the originating group. As a result, most packages have niche strengths which are different from, but complementary to, the strengths of other packages. But because the packages are separate systems, developers end up having to re-invent and implement much general functionality needed by every simulation/analysis tool. The result is duplication of effort in developing software infrastructure.

No single package currently answers all the needs of the emerging systems biology community, despite an emphasis by many developers to make their software tools omnipotent. Nor is such a scenario likely: the range of tools needed is vast, and new techniques requiring new tools are emerging far more rapidly than the rate at which any single package may be developed. For the foreseeable future, then, systems biology researchers are likely to continue using multiple packages to carry out their work. The best we can do is to develop ways to ease sharing and communication between such packages now and in the future.

These considerations lead us to believe that there is an increasingly urgent need to develop common standards and mechanisms for sharing resources within the field of systems biology. We hope to answer this need through the ERATO Systems Biology Workbench project.

THE SYSTEMS BIOLOGY MARKUP LANGUAGE

The current inability to exchange models between simulation/analysis tools has its roots in the lack of a common format for describing models. We sought to address this problem from the very beginning of the project by developing an open, extensible, model representation language.

The Systems Biology Workbench project was conceived at an ERATO-sponsored workshop held at the California Institute of Technology, USA, in December, 1999. The first meeting of all the collaborators at *The First Workshop on Software Platforms for Molecular Biology* was held at the same location in April, 2000. The participants collectively decided to begin by developing a common, XML-based (Bosak and Bray, 1999), declarative language for representing models. A draft version of this Systems Biology

Markup Language (SBML) was developed by the Caltech ERATO team and delivered to all collaborators in August, 2000. This draft version underwent extensive discussion over mailing lists and then again during *The Second Workshop on Software Platforms for Molecular Biology* held in Tokyo, Japan, November 2000. A revised version of SBML was issued by the Caltech ERATO team in December, 2000, and after further discussions over mailing lists and in meetings, a final version of the base-level definition of SBML was released publicly in March, 2001 (Hucka et al., 2001).

The Form of the Language

SBML Level 1 is the result of merging modeling-language features from the seven tools mentioned in the introduction (BioSpice, DBSolve, E-Cell, Gepasi, Jarnac, StochSim, and Virtual Cell). This base level definition of the language supports non-spatial biochemical models and the kinds of operations that are possible in these analysis/simulation tools. A number of potentially desirable features were intentionally omitted from the base language definition. Subsequent releases of SBML (termed *levels*) will add additional structures and facilities currently missing from Level 1. By freezing sets of features in SBML definitions at incremental levels, we hope to provide the community with stable standards to which software authors can design to, while at the same time allowing the simulation community to gain experience with the language definitions before introducing new elements. At the time of this writing, we are actively developing *SBML Level 2*, which is likely to include the ability to represent submodels, arrays and array connectivity, database references, three-dimensional geometry definition, and other features.

Shown at right is an example of a simple, hypothetical biochemical network that can be represented in SBML. Broken down into its constituents, this model contains a number of components: reactant species, product species, reactions, rate laws, and parameters in the rate laws. To analyze or simulate this network, additional components must be made explicit, including compartments for the species and units on the various quantities. The top level of an SBML model definition simply consists of lists of these components:

$$X_0 \xrightarrow{k_1 X_0} S_1$$
$$S_1 \xrightarrow{k_2 S_1} X_1$$
$$S_1 \xrightarrow{k_3 S_1} X_2$$

beginning of model definition
- *list of unit definitions (optional)*
- *list of compartments*
- *list of species*
- *list of parameters (optional)*
- *list of rules (optional)*
- *list of reactions*

end of model definition

The meaning of each component is as follows:

Unit definition: A name for a unit used in the expression of quantities in a model. Units may be supplied in a number of contexts in an SBML model, and it is convenient to have a facility for both setting default units and for allowing combinations of units to be given abbreviated names.

Compartment: A container of finite volume for substances. In SBML Level 1, a compartment is primarily a topological structure with a volume but no geometric qualities.

Specie: A substance or entity that takes part in a reaction. Some example species are ions such as Ca^{2++} and molecules such as glucose or ATP. The primary qualities associated with a specie in SBML Level 1 are its initial amount and the compartment in which it is located.

Parameter: A quantity that has a symbolic name. SBML Level 1 provides the ability to define parameters that are global to a model, as well as parameters that are local to a single reaction.

Reaction: A statement describing some transformation, transport or binding process that can change the amount of one or more species. For example, a reaction may describe how certain entities (reactants) are transformed into certain other entities (products). Reactions have associated rate laws describing how quickly they take place.

Rule: In SBML, a mathematical expression that is added to the differential equations constructed from the set of reactions, and can be used to set parameter values, establish constraints between quantities, etc.

A software package can read in a model expressed in SBML and translate it into its own internal format for model analysis. For instance, a package might provide the ability to simulate a model, by constructing a set of differential equations representing the network and then performing numerical time integration on the equations to explore the model's dynamic behavior. The output of the simulation might consist of plots of various quantities in the model as they change over time.

SBML allows models of arbitrary complexity to be represented. We present a simple, illustrative example of using SBML in Appendix A, but much more elaborate models are possible. The complete specification of SBML Level 1 is available from the project's World Wide Web site (`http://www.cds.caltech.edu/erato/`).

Relationships to Other Efforts

There are a number of ongoing efforts with similar goals as those of SBML. Many of them are oriented more specifically toward describing protein sequences, genes and related elements for database storage and search. These are generally not intended to be computational models, in the sense that they do not describe entities and behavioral rules in such a way that a simulation package could "run" the models.

The effort closest in spirit to SBML is CellML™ (CellML Project, 2001). CellML is an XML-based markup language designed for storing and exchanging computer-based biological models. It includes facilities for representing model structure, mathematics and additional information for database storage and search. Models are described in terms of networks of connections between discrete components; a component is a functional unit that may correspond to a physical compartment or simply a convenient modeling abstraction. Components contain variables and connections contain mappings between the variables of connected components. CellML provides facilities for grouping components and specifying the kinds of relationships that may exist between components. It uses MathML (Ausbrooks et al., 2001) for expressing mathematical relationships and provides the ability to use ECMAScript (formerly known as JavaScript; ECMA, 1999) to define functions.

The constructs in CellML tend to be at a more abstract and general level than those in SBML Level 1, and it provides somewhat more general capabilities. By contrast, SBML is closer to the internal object model used in model analysis software. Because SBML Level 1 is being developed in the context of interacting with a number of existing simulation packages, it is a more concrete language than CellML and may be better suited to its purpose of enabling interoperability with existing simulation tools. However, CellML offers viable alternative ideas and the developers of SBML and CellML are actively engaged in ensuring that the two representations can be translated between each other.

THE SYSTEMS BIOLOGY WORKBENCH

In this section, we describe how we approached the development of the Systems Biology Workbench from both philosophical and technical standpoints; we also summarize the overall architecture of the system and explain how it enables integration and sharing of software resources.

Driving Principles

The Systems Biology Workbench is primarily a system for integrating resources. It provides infrastructure that can be used to interface to software components and enable them to communicate with each other. The components in this case may be simulation codes, analysis tools, user interfaces, database interfaces, script language interpreters, or in fact any piece of software that conforms to a certain well-defined interface.

We knew from the outset that the success of the Workbench would be contingent on contributors benefitting from sharing resources through the system. For this reason, we made three commitments toward this goal:

• The Systems Biology Workbench software will be made publicly and freely available under open-source licensing (O'Reilly, 1999; Raymond, 1999). The agency funding the development of the Workbench (the Japan Science and Technology Corporation) has formally agreed that all SBW code can be placed under open-source terms. At the same time, the license terms will not force contributors to apply the same copying and distribution terms to their contributed software—developers will be free to make their components available under license terms that best suit them. They may choose to make a component available under the same open-source license, in which case it may be packaged together with the Systems Biology Workbench software distribution; however, there is nothing preventing an author from creating an SBW-compatible component that is closed-source and distributed privately.

• The Workbench architecture is designed to be symmetric with respect to facilities made available to components. All resources available through the Workbench system are equally available to all components, and no single component has a controlling share. All contributors thereby benefit equally by developing software for the Workbench.

• The direct interface between a software component and the Systems Biology Workbench is a specific application programming interface (API). The component's authors may chose to implement this API directly and publish the details of the operations provided by the component. Alternatively, they may enter into a formal agreement with us (the authors of the Workbench) in which they reveal only to us their component's API, and we will write an interface between the Workbench and this API. The latter alternative allows contributors to retain maximum confidentiality regarding their component, yet still make the component available (in binary executable form) for users of the Workbench.

The Overall Architecture of the Workbench

Although our initial focus is on enabling interaction between the seven simulation/analysis packages already mentioned, we are equally interested in creating a flexible architecture that can support future developments and new tools. We have approached this by using a combination of three key features: (1) a framework divided into layers, (2) a highly modular, extensible structure, and (3) inter-component communications facilities based on message-passing.

Layered Framework

We sought to maximize the reusability of the software that we developed for the Workbench by dividing the Workbench infrastructure into two layers: the Systems Biology Workbench itself, and a lower-level substrate called the *Biological Modeling Framework* (BMF). The latter is a general software framework that can be used in developing a variety of biological modeling and analysis software applications. It not directly tied to the current architecture of SBW, allowing us the freedom to evolve and change SBW in the future while still maintaining a relatively stable foundation.

BMF provides basic scaffolding supporting a modular, extensible application architecture (see below), as well as a set of useful software components that can be used as black boxes in constructing a system (cf. Fayad et al., 1999). Other projects should be able to start with BMF, add their own domain- and task-specific elements, and thereby implement a system specialized for other purposes. This is how the neuroscience-oriented Modeler's Workspace (Hucka et al, 2000) is being implemented. Computational biologists and other users do not need to be aware of the existence of BMF—it is scaffolding used by the developers of SBW and other tools, and not a user-level construct.

SBW is a particular collection of application-specific components layered on top of BMF. These collectively implement what users experience as the "Systems Biology Workbench". Some components add functionality supporting the overall operation of the Workbench, such as the message-passing communications facility; other components implement the interfaces to the specific simulation/analysis tools made available in the Workbench. Figure 6.1 illustrates the overall organization of the layers.

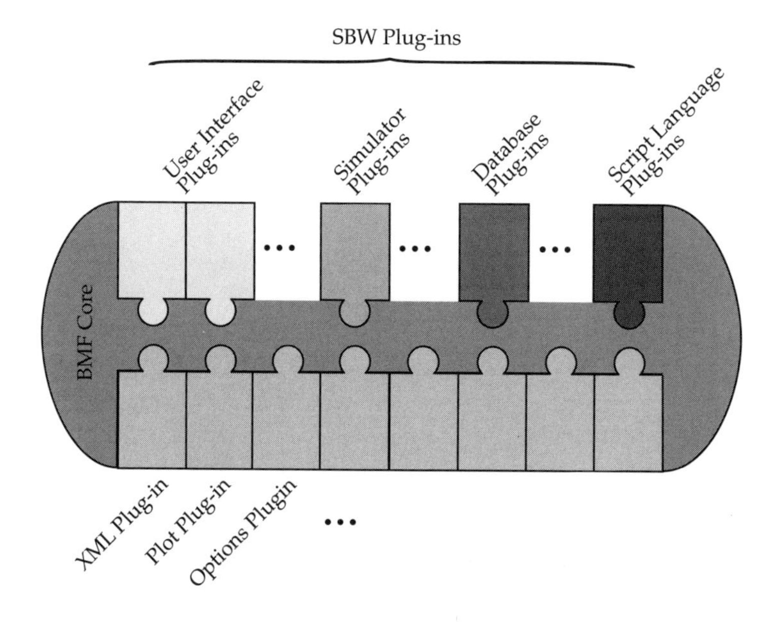

Figure 6.1 The Systems Biology Workbench (SBW) is a collection of software components layered on top of a simple plug-in architecture called the Biological Modeling Framework (BMF).

Highly Modular, Extensible Architecture

The Biological Modeling Framework layer that underlies SBW is implemented in Java and provides (1) support for managing pluggable components ("plug-ins") and (2) a collection of basic components (such as an XML Schema-aware parser, file utilities, basic plotting and graphing utilities, etc.) that are useful when implementing the typical kinds of applications for which BMF is intended.

The kinds of application-specific plug-ins that comprise SBW can generally be grouped by their purposes: user interfaces, simulator/analyzer interfaces, scripting language interfaces, and database interfaces. Other kinds are possible, but these are the primary application-specific plug-in types that we foresee being developed. There can be any number of plug-ins in a given system, subject to the usual limitations on computer resources such as memory. Each plug-in needs to conform to certain minimal interface requirements dictated by the framework, discussed further below. A plug-in can make use of any public services provided by BMF, the core plug-ins, and application-specific plug-ins.

By virtue of the software environment provided by the Java 2 Platform (Flanagan, 1999; Gosling et al., 1996), plug-ins can be loaded dynamically, without recompiling or even necessarily restarting a running application. This can be used to great advantage for a flexible and powerful environment. For example, an application could be designed to be smart about how it handles data types, loading specialized plug-ins to allow a user to interact with particular data objects on an as-needed basis. If the user does not already have a copy of a certain plug-in stored on the local disk, the plug-in could be obtained over the Internet, much like current-generation Web browsers can load plug-ins on demand. In this manner, plug-ins for tasks such as displaying specific data types or accessing third-party remote databases could be easily distributed to users.

Message-Passing for Inter-Component Communications

One of the challenges in developing a modular system, especially one that allows incremental addition and change by different developers, is designing appropriate interfaces between components. Knowledge of an element's interface necessarily becomes encoded in its structure; otherwise, component *A* would not know how to communicate with component *B*. Many frameworks are designed around a hierarchy of object classes and interfaces; this lets them provide a rich selection of features. However, for this project, class hierarchies have two important disadvantages:

- Methods and calling conventions for accessing different objects become scattered throughout the structure of each component in the system. The effects of changing an interface are not localized in client code: changing the interface of a fundamental object may require rewriting code in many different locations in every other component that uses it.

• The task of creating interfaces to components written in other programming languages is more complex. If the component is a library, the foreign-function bridge (whether it is implemented using the Java Native Interface [JNI] or other) must expose all of the methods provided by the component's interface, which requires substantial programming effort and significant maintenance. Similarly, if the component is a stand-alone application, the mechanism used to communicate with it must provide access to all or most of the classes and methods provided by the component's interface. CORBA (Object Management Group, 2001; Seetharman, 1998; Vinoski, 1997) is one technology that could be used to cope with these issues, but we decided to avoid requiring its use in SBW because we feared its complexity would be too daunting to potential contributors.

We began developing SBW using the common approach of designing object class hierarchies representing different functions and simulation/analysis capabilities, but soon decided that the problems above would become too onerous. We chose instead to base inter-component communications on passing messages via byte streams.

In this approach, each component or component wrapper needs to expose a simple programmatic interface, consisting of only a handful of methods. The main method in this interface (`PluginReceive`) is the entry point for exchanging messages. Other methods in the interface provide a way for starting the execution of the component (`PluginStart`), and for obtaining its name and a list of strings describing the kinds of capabilities it implements (`PluginRegistrationInformation`). The latter can be used by other components to discover programmatically what services a new component provides.

A *message* in this framework is a stream of bytes that encodes a service identifier and a list of arguments. The service identifier is determined from the list of services advertised by the component. The arguments are determined by the particular service. For example, a command to perform steady-state analysis on a biochemical network model would require a message that encodes the model and a list of parameters on the kind of analysis desired. The result passed back by the component would be in the form of another message.

The representation of the data in a message is encoded according to a specific scheme. The scheme in SBW allows for the most common data types to be packaged into a message. Each element in a message is preceded by a byte that identifies its data type. The types currently supported include character strings, integers, double-sized floating-point numbers, arrays of homogeneous elements, and lists of heterogeneous elements.

How does this approach help cope with the two problems listed above? At first glance, it may seem that this approach merely hides functionality behind a simple façade. After all, changing the operation of a component still requires other components to be changed as well, for example to compose and parse messages differently as needed. However, in this approach, the effects of actual interface changes are more *localized*, typically affecting fewer objects in other components.

The message-passing approach also simplifies the task of interfacing to components implemented in different programming languages. Rather than have to provide native-code interfaces (say, using Java JNI) to every method in a large class hierarchy, only a few methods must be implemented. Likewise, it is much simpler to link components that run as separate processes or on remote computers. A simple message-passing stream is easily implemented through a TCP/IP socket interface, the simplest and most familiar networking facility available nearly universally on almost every computer platform.

The current message-passing scheme can be used to exchange messages encoded in XML, which makes this approach similar to XML-RPC (Winer, 1999) and SOAP (Box et al., 2000). However, our message protocol allows other data types to be encoded as well. Using XML exclusively would require binary data to be encoded into, and unencoded from, a textual representation, which would impact performance and potentially affect floating-point accuracy. We therefore designed the protocol to allow binary data as well as XML to be exchanged.

Advantages of an Extensible Framework Approach

The modular framework approach is pervasive throughout the design of the system. Both the underlying BMF layer and an application layer such as SBW are implemented as plug-ins attached to a small core. Nearly all of the functionality of both layers are determined by the plug-ins themselves.

The primary benefits of using a modular framework approach accrue to software developers. For a developer who would like to build upon BMF and create a new system, or take an existing system such as SBW and create enhancements or specialized components, the following are some of the benefits (Fayad et al., 1999):

- *Modularity*. A framework is composed of modules, each of which encapsulates implementation details behind a stable interface. Design and implementation changes to individual modules are less likely to have an impact on other modules or the overall system.
- *Reusability*. A framework offers a set of elements that represent common solutions to recurring problems. Reusing the elements saves design and development time.
- *Extensibility*. Extensibility is designed into a framework by providing well-defined ways to add new modules and extend existing modules. In its most essential form, a framework is a substrate for bringing software modules together.

Although frameworks were invented by software developers to simplify the implementation of complex software, users also benefit when a system is based on a framework approach. For a biologist or other user who would like to employ a tool built on top of BMF and SBW, there are two primary gains:

- *Control.* Users are given greater control over the composition of a framework-based system than a system based on a more traditional architecture. They can use just those modules that they need, and they have the freedom to chose different implementations of the same functionality, or even develop their own implementations, all without altering the rest of the system.
- *Reusability.* A successful framework may be reused to implement other domain-specific tools, reducing the burden on a user by allowing them to carry over their experiences involving the common parts.

Motivations for Using Java

We chose Java as the implementation language for the underlying BMF layer of SBW because it offers a number of attractive features and meets several objectives. In particular, Java arguably provides one of the most portable cross-platform environments currently available. Java also provides a built-in mechanism for dynamic loading of code, simplifying the implementation of an architecture oriented around plug-ins. Finally, Java provides a rich platform for development, with such things as remote invocation facilities and GUI widgets, on all supported platforms.

It is worth noting that plug-ins for the system are *not required* to be written in Java. Java provides mechanisms for interfacing to software written in other languages, through the Java Native Interface. Thus, although Java is used to implement the core of the system, plug-ins can be written in other languages and incorporated into an application built on top of the framework.

Although Java has received negative publicity in the past with respect to performance (Tyma, 1998), we do not feel that choosing Java will have a significant impact on run-time performance. The reason is that the core of the Systems Biology Workbench is a thin layer and most of the execution time in an application is spent within application-specific plug-ins. Those can be written in other languages if performance becomes an issue.

SUMMARY AND STATUS

The aim of the Systems Biology Workbench project is to create a modular, open software environment that enables different simulation and analysis tools to be used together for systems biology research. As part of this effort, we have also developed a model description language, the Systems

Biology Markup Language, that can be used to represent models in a form independent of any specific simulation/analysis tool. SBML is based on XML for maximal flexibility, interchangeability, and future compatibility.

Availability

We will make the software available under open-source terms from the Caltech ERATO team's web site, `http://www.cds.caltech.edu/erato/`. At the time of this writing, we are in the process of developing and implementing the core functionality of the Systems Biology Workbench, along with an initial set of plug-ins. The aim of this effort is to demonstrate the concepts described above and provide a medium through which we will develop and refine the APIs. We expect to make this initial implementation available in the first half of 2001, and to release the first full version of the Workbench by the end of 2001.

Future Plans

The final specification for SBML Level 1 was released in March, 2001. The relevant documents are available from the Caltech ERATO team's web site, mentioned above. SBML Level 2 is currently under development, and we anticipate making a preliminary specification available later in the year 2001. We will publish the specification documents on the web site as they become available.

ACKNOWLEDGMENTS

We are grateful for comments, advice and help from fellow ERATO Kitano Systems Biology project members Mark Borisuk, Mineo Morohashi, Eric Mjolsness, Bruce Shapiro, and Tau-Mu Yi.

SBML Level 1 was developed with the help of many people. We wish to acknowledge in particular the authors of BioSpice, DBSolve, E-Cell, Gepasi, StochSim, and Virtual Cell, and members of the `sysbio` mailing list. We are especially grateful to the following people for discussions and knowledge: Dennis Bray, Athel Cornish-Bowden, David Fell, Carl Firth, Warren Hedley, Martin Ginkel, Igor Goryanin, Jay Kaserger, Andreas Kremling, Nicolas Le Novère, Les Loew, Daniel Lucio, Pedro Mendes, Eric Mjolsness, Jim Schaff, Bruce Shapiro, Tom Shimizu, Hugh Spence, Joerg Stelling, Kouichi Takahashi, Masaru Tomita, and John Wagner.

APPENDIX

A EXAMPLE OF A MODEL ENCODED IN XML USING SBML

Consider the following hypothetical branched system:

$$X_0 \quad \xrightarrow{k_1 X_0} \quad S_1$$

$$S_1 \quad \xrightarrow{k_2 S_1} \quad X_1$$

$$S_1 \quad \xrightarrow{k_3 S_1} \quad X_2$$

The following is the main portion of an XML document that encodes the model shown above:

```
<sbml level="1" version="1">
    <model name="Branched">
        <notes>
            <body xmlns="http://www.w3.org/1999/xhtml">
                <p>Simple branched system.</p>
                <p>reaction-1:   X0 -> S1; k1*X0;</p>
                <p>reaction-2:   S1 -> X1; k2*S1;</p>
                <p>reaction-3:   S1 -> X2; k3*S1;</p>
            </body>
        </notes>
        <listOfCompartments>
            <compartment name="A" volume="1"/>
        </listOfCompartments>
        <listOfSpecies>
            <specie name="S1" initialAmount="0" compartment="A"
                    boundaryCondition="false"/>
            <specie name="X0" initialAmount="0" compartment="A"
                    boundaryCondition="true"/>
            <specie name="X1" initialAmount="0" compartment="A"
                    boundaryCondition="true"/>
            <specie name="X2" initialAmount="0" compartment="A"
                    boundaryCondition="true"/>
        </listOfSpecies>
        <listOfReactions>
            <reaction name="reaction_1" reversible="false">
                <listOfReactants>
                    <specieReference specie="X0"
                                     stoichiometry="1"/>
                </listOfReactants>
                <listOfProducts>
                    <specieReference specie="S1"
                                     stoichiometry="1"/>
                </listOfProducts>
                <kineticLaw formula="k1 * X0">
                    <listOfParameters>
                        <parameter name="k1" value="0"/>
                    </listOfParameters>
                </kineticLaw>
            </reaction>
            <reaction name="reaction_2" reversible="false">
                <listOfReactants>
                    <specieReference specie="S1"
                                     stoichiometry="1"/>
```

```
                </listOfReactants>
                <listOfProducts>
                    <specieReference specie="X1"
                                     stoichiometry="1"/>
                </listOfProducts>
                <kineticLaw formula="k2 * S1">
                    <listOfParameters>
                        <parameter name="k2" value="0"/>
                    </listOfParameters>
                </kineticLaw>
            </reaction>
            <reaction name="reaction_3" reversible="false">
                <listOfReactants>
                    <specieReference specie="S1"
                                     stoichiometry="1"/>
                </listOfReactants>
                <listOfProducts>
                    <specieReference specie="X2"
                                     stoichiometry="1"/>
                </listOfProducts>
                <kineticLaw formula="k3 * S1">
                    <listOfParameters>
                        <parameter name="k3" value="0"/>
                    </listOfParameters>
                </kineticLaw>
            </reaction>
        </listOfReactions>
    </model>
</sbml>
```

The XML encoding shown above is quite straightforward. The outermost container is a tag, `smbl`, that identifies the contents as being systems biology markup language. The attributes `level` and `version` indicate that the content is formatted according to version 1 of the Level 1 definition of SBML. The `version` attribute is present in case SBML Level 1 must be revised in the future to correct errors.

The next-inner container is a single `model` element that serves as the highest-level object in the model. The model has a name, "Branched". The model contains one compartment, four species, and three reactions. The elements in the `listOfReactants` and `listOfProducts` in each reaction refer to the names of elements listed in the `listOfSpecies`. The correspondences between the various elements should be fairly obvious.

The model includes a `notes` annotation that summarizes the model in text form, with formatting based on XHTML. This might be useful for a software package that is able to read such annotations and render them in HTML.

References

Abbott, A. (1999). Alliance of US Labs Plans to Build Map of Cell Signalling Pathways. *Nature* 402:219–220.

Arkin, A. P. (2001). *Simulac* and *Deduce*. Available via the World Wide Web at `http://gobi.lbl.gov/~aparkin/Stuff/Software.html`.

Ausbrooks, R., Buswell, S., Dalmas, S., Devitt, S., Diaz, A., Hunter, R., Smith, B., Soiffer, N., Sutor, R., and S. Watt. (2001). Mathematical Markup Language (MathML) Version 2.0: W3C Proposed Recommendation 08 January 2001. Available via the World Wide Web at `http://www.w3.org/TR/2001/PR-MathML2-20010108/`.

Bosak, J., and Bray, T. (1999). XML and the Second-Generation Web. *Scientific American*, May. Also available via the World Wide Web at `http://www.sciam.com/1999/0599issue/0599bosak.html`.

Box, D., Ehnebuske, D., Kakivaya, G., Layman, A., Mendelsohn, N., Nielsen, H. F., Thatte, S., and Winer, D. (2000). Simple Object Access Protocol (SOAP) 1.1: W3C Note 08 May 2000. Available via the World Wide Web at `http://www.w3.org/TR/SOAP/`.

Bray, D., Firth, C., Le Novère, N., and Shimizu, T. (2001). *StochSim*, Available via the World Wide Web at `http://www.zoo.cam.ac.uk/comp-cell/StochSim.html`.

CellML Project. (2001). CellML Project Home Page. `http://www.cellml.org/`.

ECMA. (1999). ECMAScript Language Specification: Standard ECMA-262, 3rd Edition. Available via the World Wide Web at `http://www.ecma.ch/ecma1/STAND/ECMA-262.HTM`.

Gosling, J., Joy, B., and Steele, G. (1996). The Java™ Language Specification. Addison-Wesley.

Fayad, M. E., Schmidt, D. C., and Johnson, R. E. (1999). *Building Application Frameworks*. Wiley.

Flanagan, D. (1999). *Java in a Nutshell*. O'Reilly & Associates.

Gilman, A. (2000). A Letter to the Signaling Community. Alliance for Cellular Signaling, The University of Texas Southwestern Medical Center. Available via the World Wide Web at `http://afcs.swmed.edu/afcs/Letter_to_community.htm`.

Goryanin, I. (2001). DBsolve: Software for Metabolic, Enzymatic and Receptor-Ligand Binding Simulation. Available via the World Wide Web at `http://websites.ntl.com/~igor.goryanin/`.

Goryanin, I., Hodgman, T. C., and Selkov, E. (1999). Mathematical Simulation and Analysis of Cellular Metabolism and Regulation. *Bioinformatics* 15(9):749–758.

Hucka, M., Beeman, D., Shankar, K., Emardson, S., and Bower, J. M. (2000). The Modeler's Workspace: A Tool for Computational Neuroscientists. *Society for Neuroscience Abstracts* 30, 21.69.

Hucka, M., Finney, A., Sauro, H. S., and Bolouri, H. (2001). Systems Biology Markup Language (SBML) Level 1: Structures and Facilities for Basic Model Definitions. Available via the World Wide Web at `http://www.cds.caltech.edu/erato`.

Mendes, P. (1997). Biochemistry by Numbers: Simulation of Biochemical Pathways with Gepasi 3. *Trends in Biochemical Sciences* 22:361–363.

Mendes, P. (2001). *Gepasi 3.21*. Available via the World Wide Web at `http://www.gepasi.org/`.

Morton-Firth, C. J., and Bray, D. (1998). Predicting Temporal Fluctuations in an Intracellular Signalling Pathway. *Journal of Theoretical Biology* 192:117–128.

O'Reilly, T. (1999). Lessons from Open-Source Software Development. *Communications of the ACM* 42(4):32–37.

Object Management Group. (2001). *CORBA/IIOP 2.3.1*. Specification documents available via the World Wide Web at `http://www.omg.org/`.

Popel, A., and Winslow, R. L. (1998). A Letter From the Directors ... Center for Computational Medicine & Biology, Johns Hopkins School of Medicine, Johns Hopkins University. Available via the World Wide Web at `http://www.bme.jhu.edu/ccmb/ccmbletter.html`.

Raymond, E. S. (1999). *The Cathedral & the Bazaar*, O'Reilly & Associates. (This is the book form; the paper alone is available via the World Wide Web at `http://www.tuxedo.org/~esr/writings/cathedral-bazaar/`.)

Schaff, J., Slepchenko, B., and Loew, L. M. (2000). Physiological Modeling with the Virtual Cell Framework. In *Methods in Enzymology*, Academic Press, 321:1–23.

Schaff, J., Slepchenko, B., Morgan, F., Wagner, J., Resasco, D., Shin, D., Choi, Y. S., Loew, L., Carson, J., Cowan, A., Moraru, I., Watras, J., Teraski, M., and Fink., C. (2001). *Virtual Cell*. Available over the World Wide Web at `http://www.nrcam.uchc.edu/`.

Seetharaman, K. (1998). The CORBA Connection. *Communications of the ACM* 41(10):34–36.

Smaglik, P. (2000). For My Next Trick ... *Nature* 407:828–829.

Sauro, H. M. (1991). SCAMP: A Metabolic Simulator and Control Analysis Program. *Mathl. Comput. Modelling* 15:15–28.

Sauro, H. M. and Fell, D. A. (2000). Jarnac: A System for Interactive Metabolic Analysis. In Hofmeyr, J-H. S., Rohwer, J. M., and Snoep, J. L. (eds.), *Animating the Cellular Map 9th International BioThermoKinetics Meeting*, Stellenbosch University Press.

Tomita, M., Hashimoto, K., Takahashi, K., Shimizu, T., Matsuzaki, Y., Miyoshi, F., Saito, K., Tanida, S., Yugi, K., Venter, J. C., and Hutchison, C. (1999). E-CELL: Software Environment for Whole Cell Simulation. *Bioinformatics* 15(1):72–84.

Tomita, M., Nakayama, Y., Naito, Y., Shimizu, T., Hashimoto, K., Takahashi, K., Matsuzaki, Y., Yugi, K., Miyoshi, F., Saito, Y., Kuroki, A., Ishida, T., Iwata, T., Yoneda, M., Kita, M., Yamada, Y., Wang, E., Seno, S., Okayama, M., Kinoshita, A., Fujita, Y., Matsuo, R., Yanagihara, T., Watari, D., Ishinabe, S., and Miyamoto, S. (2001). *E-Cell*. Available via the World Wide Web at `http://www.e-cell.org`.

Tyma, P. (1998). Why Are We Using Java Again? *Communications of the ACM* 41(6):38–42.

Vinoski, S. (1997). CORBA: Integrating Diverse Applications Within Distributed Heterogeneous Environments. *IEEE Communications Magazine*. Feb.

Winer, D. (1999). XML-RPC Specification. Available via the World Wide Web at `http://www.xml-rpc.com/spec/`.

7 Automatic Model Generation for Signal Transduction with Applications to MAP-Kinase Pathways

Bruce E. Shapiro, Andre Levchenko, and Eric Mjolsness

We describe a general approach to automatic model generation in the description of dynamic regulatory networks. Several potential areas of application of this technique are outlined. We then describe how a particular implementation of this approach, Cellerator, has been used to study the mitogen-activated protein kinase (MAPK) cascade. These signal transduction modules occur both in solution and when bound to a scaffold protein, and we have generalized the technique to include both types of module. We show that the results of simulations with the Cellerator–created model are consistent with our previously published report, where an independently written model was developed. New results made possible by the use of Cellerator are also presented. An important aspect of Cellerator operation – explicit output description at several steps during model generation – is emphasized. This design allows intervention and modification of the model "on the go" leading to both a more flexible of model description and a straightforward error correction mechanism. We also outline our future plans in Cellerator development.

INTRODUCTION

In the past few decades the rapid gain of information about intracellular signal transduction and genetic networks has led to the view of regulatory biomolecular circuits as highly structured multi-component systems that have evolved to perform optimally in very uncertain environments. This emergent complexity of biochemical regulation necessitates the development of new tools for analysis, most notably computer assisted mathematical models. Computer modeling has proved to be of crucial importance in the analysis of genomic DNA sequences and molecular dynamics simulations and is likely to become an indispensable tool in biochemical and genetic research. Several platforms have been (or are being) developed that enable biologists to do complex computational simulations of various aspects of cellular signaling and gene regulatory networks.

In spite of their promise, these new modeling environments have not been widely utilized in the biological research community. Arguably, among the reasons for this is a relative inaccessibility of the modeling interface for the typical classically trained geneticist or biochemist. Instead of cartoon representations of signaling pathways, in which activation can be represented simply by an arrow connecting two molecular species, users are often asked to write specific differential equations or chose among different modeling approximations. Even for fairly modest biomolecular circuits such a technique would involve explicitly writing dozens (or even hundreds) of differential equations, a job that can be tedious, difficult, and highly error prone, even for an experienced modeler. Thus it would be extremely helpful to have a modeling interface that would automatically convert a cartoon- or reaction-based biochemical pathway description into a mathematical representation suitable for the solvers built into various currently existing software packages.

In addition to being more accessible to a wider research community, a tool allowing the automatic generation of mathematical models would facilitate the modeling of complex networks and interactions. For example, in intracellular signal transduction it is not uncommon to find multi-molecular complexes of modifiable proteins. The number of different states that a multi-molecular complex, along with the number of equations required to fully describe the dynamics of such a system, increases exponentially with the number of participating molecules or classes of molecules. One typical complex – scaffolds in MAPK cascades – will be studied in detail later in this report. It is often the case that the dynamics of each state is of interest. A modeler then faces the unpleasant, and potentially error prone task, of writing dozens, if not hundreds, of equations. Automatic equation generation can significantly ease this task.

In this report we consider a general approach to automatic model generation for the description of dynamic regulatory networks. Several potential areas of application of this technique will be outlined. We then will describe how a particular implementation of this approach, Cellerator, has been used to study the mitogen-activated protein kinase (MAPK) cascade signal transduction modules operating in solution or when bound to a scaffold protein. An important aspect of Cellerator operation – explicit output description and flexible user intervention at several steps through the model generation – will be emphasized. This design, which allows intervention and modification of the model "on the fly" leads to increased model design flexibility and provides an immediate error correction mechanism.

Canonical Forms for Cell Simulation

We can loosely classify the components needed to perform cell simulation in order of their biological complexity: simple chemical reactions including degradation, enzymatic reactions in solution, multi-molecular complexes with a non-trivial number of states (e.g., scaffold proteins), multiple interacting and non-overlapping pathways, transcription, translation, intracellular components, transport processes and morphogenesis. We will examine these processes and attempt to derive general *canonical forms* that can be used to describe these processes in the following paragraphs. These canonical forms can be either *input forms*, such as chemical reactions, or *output forms*, such as differential equations that are automatically generated by the program. It is crucial to identify these canonical forms so that an efficient mapping from the input forms to the output forms can be implemented. Specific examples of how these forms may be implemented in a computer program are given in the following section.

Biochemistry is frequently referred to as the language of biology, in much the same way that mathematics has been called the language of physics. Cellular activity is generally expressed in terms of the biochemical cascades that occur. These chemical reactions constitute the core of our input forms; the corresponding differential equations constitute the core of our output forms. (Differential equations can be thought of as output because they are passed on to solver and/or optimizer modules to handle). A fundamental library of simple chemical reactions can be quickly developed; such reactions take the form

$$\sum_{X_i \in S' \subset S} X_i \xrightarrow{k} \sum_{Y_i \in S'' \subset S} Y_i \tag{7.1}$$

where S is a set of reactants and S' and S'' are (possible empty and possibly non-distinct) subsets of S and k is a representation of the rate at which the reaction proceeds. In general there are rarely more than two elements in either S' or S'' but it is possible for there to be more. For example, all of the following chemical reactions fall into this form:

$$\begin{aligned}
&A + B \rightarrow C = AB && \text{complex formation}\\
&C = AB \rightarrow A + B && \text{dissociation}\\
&A \rightarrow B && \text{conversion}\\
&A \rightarrow \phi && \text{degradation}\\
&\phi \rightarrow A && \text{creation (}e.g.\text{, through transcription)}
\end{aligned} \tag{7.2}$$

Enzyme kinetic reactions, which are usually written as

$$S + E \rightarrow P + E \tag{7.3}$$

where E is an enzyme that facilitates the conversion of the substrate S into the product P, would also fall into this class. More generally, equation (7.3) is a simplification of the cascade

$$S + E \leftrightarrow SE \rightarrow S + P \tag{7.4}$$

where the bi-directional arrow indicates that the first reaction is reversible. Thus (7.4) is equivalent to the triplet of reactions

$$\{S + E \rightarrow SE, SE \rightarrow S + E, SE \rightarrow S + P\} \tag{7.5}$$

The reactions (7.4) or (7.5) can be written compactly with the following double-arrow notation

$$S \overset{E}{\Longrightarrow} P \tag{7.6}$$

which should be read as "the conversion of S to P is catalyzed by an enzyme E." If there is also an second enzyme, G, that can catalyze the reverse reaction

$$P \overset{G}{\Longrightarrow} S = \{P + G \rightarrow GP, GP \rightarrow G + P, GP \rightarrow G + S\} \tag{7.7}$$

we further use the double-double arrow notation

$$S \underset{G}{\overset{E}{\Longleftrightarrow}} P \tag{7.8}$$

to compactly indicate the pair of enzymatic reactions given by (7.6) and (7.7). The enzyme above the arrow always facilitates the forward reaction, and the enzyme beneath the reaction always facilitates the reverse reaction. For example, E might be a kinase and G might be a phosphatase molecule. Since each of equations (7.6) and (7.7) represent a triplet of simpler reactions, we observe that the notation of equation (7.8) compactly represents a total of six elementary reactions, each of which is in the form given by equation (7.1). We therefore take equation (7.1) as our input canonical form for chemical reactions. The corresponding output canonical form is given by the set of differential equations

$$\tau_i \dot{X}_i = \sum_\alpha c_{i\alpha} \prod_j X_j^{n_{i\alpha j}} \tag{7.9}$$

where the τ_i and $c_{i\alpha}$ are constants that are related to the rate constants , the signs of the c_{ia} are determined from which side of equation (7.1) the terms in equation (7.9) correspond to, and the $n_{i\alpha j}$ represent the cooperativity of the reaction. The summation is taken over all equations in which X_i appears. Multi-molecular reactions (e.g., binding to a scaffold protein) and multiple interacting and overlapping pathways are described in much the same way - there are just more reactions that must be included in our model. The canonical forms (7.1) and (7.9) can still describe each one of these reactions.

Genetic transcription and translation into proteins can be described by an extension of equation (7.9) to include terms of the form

$$\tau_i \dot{X}_i = \prod_\beta \frac{c_{i\beta} X_\beta^{n_\beta}}{K_{i\beta}^{n_\beta} + X_\beta^{n_\beta}} \tag{7.10}$$

where the product runs over the various transcription factors X_β that influence production of X_i. If there are any reactions of the form (7.1) for X_i then the expression on the right side of equation (7.10) would be added to the right hand side of (7.9). In a more realistic system, a gene would be influenced by a (possibly large) set of promoter and enhancer elements X_i that bind to different sites. A hierarchical model could describe this set of interactions

$$\tau_i \dot{X}_i = \frac{J u_i}{1 + J u_i} - \lambda_i X_i \tag{7.11}$$

$$u_i = \prod_{\alpha \in i} \frac{1 + J_\alpha \tilde{v}_\alpha}{1 + \hat{J}_\alpha \tilde{v}_\alpha} \tag{7.12}$$

$$\tilde{v}_\alpha = \frac{\tilde{K}_\alpha \tilde{u}_\alpha}{1 + \tilde{K}_\alpha \tilde{u}_\alpha} \tag{7.13}$$

$$\tilde{u}_\alpha = \prod_{b \in \alpha} \frac{1 + K_b v_{j(b)}^{n(b)}}{1 + \hat{K}_b v_{j(b)}^{n(b)}} \tag{7.14}$$

where i and j index transcription factors, α indexes promoter modules, b indexes binding sites, the function $j(b)$ determines which transcription factor j binds at site j, the J and K are constants, and λ is a degradation rate.

Sub-cellular components represent a higher order of biological complexity. If we assume perfect mixing each component can be treated as a separate pool of reactants which we can describe by the reaction

$$X_A \rightarrow X_B \tag{7.15}$$

This is taken to mean that X in pool A is transported into pool B at some rate. When the concentration changes and distances involved are small such processes can be described by the canonical forms in equation (7.1). In large or elongated cells with long processes (such as neurons) or when the molecules have a net charge the transport process defined in equation (7.15) can not be described by the output canonical form (7.9). Instead we must modify this ordinary differential equation into a partial differential equation to allow for diffusion,

$$\tau_i \frac{\partial X_i}{\partial t} = \nabla \bullet (D_i \nabla X_i + C_i D_i \nabla V) + \sum_{\alpha} c_{i\alpha} \prod_{j} X_j^{n_{i\alpha j}} \tag{7.16}$$

where the D_i are (possibly spatially dependent) diffusion constants for species X_i, C_i are charge and temperature dependent constants, and V is the voltage. Other voltage and pressure dependent movement between compartments (especially those with membranes) that are controlled by channels and transport proteins could be described by including additional terms on the right hand side of equation (7.16) (e.g., Hodgkin-Huxley type expressions).

IMPLEMENTATION

In standard biochemical notation, protein cascades are represented by a arrow-sequence of the form

$$A \Rightarrow B \Rightarrow \cdots \tag{7.17}$$

where each step (the A, B,...) would represent, for example, the activation of a particular molecular species. Our goal is to translate the cascade (7.17) into a computable form while retaining the biological notation in the user interface. Mathematically, we can specify such as cascade as a multiset

$$C = \{P, R, IC, I, F\} \tag{7.18}$$

where P is a set of proteins, R is a set of reactions, IC is a set of initial conditions, I is a set of input functions, and F is a set of output functions.

To illustrate this transformation process (from the biochemical notation, such as in equation (7.17), to the mathematical notation, as in equation (7.18)), we consider the example where equation (7.17) represents a simple linear phosphorylation cascade. In this case equation (7.17) would mean that A facilitates the phosphorylation of B, which in turn facilitates the phosphorylation of C, and so forth. In general, a cascade can have any length, so we define the elements of a cascade with a simple indexed notation, *e.g.*,

$$K_4 \Rightarrow K_3 \Rightarrow K_2 \Rightarrow K_1 \tag{7.19}$$

where K is used to indicate that all the members of the cascade induce phosphorylation of their substrates, that is they are kinases. In general, activation can proceed by any specified means.

This indexed notation is always used internally by the program. The user, however, has the option of using either common names or the indexed variables. There is still a great deal of information hidden in this expression, such as how many phosphate groups must be added to make each successive protein active. In the MAPK cascade for example (as explained below), the input signal that starts this cascade is K_4. The output, however, is not K_1, as this notation would suggest, but a doubly phosphorylated version of K_1. Hence for MAPK cascade we introduce a modified notation:

$$\begin{array}{lllll} K_3 & \overset{K_4}{\Longrightarrow} & K_3^* & & \\ K_2 & \overset{K_3^*}{\Longrightarrow} & K_2^* & \overset{K_3^*}{\Longrightarrow} & K_2^{**} \\ K_1 & \overset{K_2^{**}}{\Longrightarrow} & K_1^* & \overset{K_2^{**}}{\Longrightarrow} & K_1^{**} \end{array} \tag{7.20}$$

where each phosphate group that has been added is indicated with an asterisk. From this notation it is clear that the input is K_4 and the output is K_1^{**}.

In general, suppose we have a cascade formed by n proteins K_1, K_2,..., K_n, and that the ith protein K_i can be phosphorylated a_i times. Denote by K_i^j the fact that kinase K_i has be phosphorylated j (possibly zero) times. The set P of all kinases ${K_i}^j$ in an n-component cascade is then

$$P = \left\{ K_i^j | i = 1, 2, ..., n, \ j = 0, 1, \ldots, a_i \right\} \tag{7.21}$$

The *reactions* in the cascade are of the form

$$R = \left\{ K_i^j \overset{K_{i+1}^{a_{i+1}}}{\Longrightarrow} K_i^{j+1} \mid i = 1, \ldots, n-1, \ j = 0, \ldots, a_j - 1 \right\} \tag{7.22}$$

We note at this point that this notation describes a linear cascade, in which each element K_i is only phosphorylated by the active form of K_{i+1}. It does not include other reactions, when, for example, K_3 might, under special circumstances, phosphorylate K_1 directly without the intermediate step of first phosphorylating K_2. Such additional reactions could be added, but they have been omitted from this presentation to simplify the discussion. We can also add the dephosphorylation enzymes, or phosphatases, with a double-arrow notation:

$$R = \left\{ K_i^j \underset{Ph_i}{\overset{K_{i+1}^{a_{i+1}}}{\Longrightarrow}} K_i^{j+1} \mid i = 1, \ldots, n-1, \ j = 0, \ldots, a_j - 1 \right\} \tag{7.23}$$

In general, it is not necessary to specify explicit conservation laws with this notation because they are built directly into the equations. For example, we do not have to separately specify that the quantities

$$K_i^{Total} = \sum_{j=0}^{a_i} K_i^j \tag{7.24}$$

because this is implicit in the differential equations that are built using this notations. We do, however, have to specify the initial conditions,

$$IC = \left\{ K_i^j(0) \mid i = 1, 2, ..., n, \ j = 0, 1, \ldots, a_i \right\} \tag{7.25}$$

Next, we need to specify how the cascade is initiated. For example if K_4 is not present until some time t_{on} and then is fixed at a level c, would write the set of *input functions* as

$$I = \{K_4(t) = cH(t - t_{on})\} \quad \text{where} \quad H(t) = \begin{cases} 0 & (t < 0) \\ 1 & (t \geq 0) \end{cases} \tag{7.26}$$

is the Heaviside step function. In some cases, we are only interested in the total quantity of each substance produced as a function of time, e.g., $K_i^j(t)$. More generally, we would also specify a set of *output functions* F. For example we might have $F = \{ f, g \}$ where $f(T)$ is the total accumulated protein concentration after some time T,

$$f(T) = \int_{t_{on}}^{T} K_1^{a_1}(t)dt \tag{7.27}$$

and $g(c)$ is the steady state concentration of activated kinase,

$$g(c) = \lim_{t_{on} \to \infty} \left[\lim_{t \to \infty} K_1^{a_1}(t) \right] \tag{7.28}$$

where c is the input signal specified I. Then the cascade is then completely specified by the multiset $C = \{P, R, IC, I, F\}$.

If we have an additional regulatory protein, such as a scaffold that holds the various proteins in equation (7.20) together there are additional reactions. These describe binding of the enzymes to the scaffold and phosphorylation within the scaffold. We describe the scaffold itself by defining an object $S_{p_1, p_2, \cdots, p_n}$ where n is as before (the number of kinases that may bind to the scaffold, or alternatively, the number of "slots" in the scaffold) and $p_i \in \varepsilon, 0, 1, \ldots, a_i$ indicates the state of phosphorylation of the proteins in each slot. Thus if $p_i = \varepsilon$ (or, alternatively, -1) the slot for K_i is empty, if $p_i = 0$, K_i^0 is in the slot, etc. For a three-slot scaffold, for example, we would add to the set P the following set

$$P' = \left\{ S_{ijk} \;\middle|\; i = \varepsilon, 0, 1, \ldots, a_1, \;\; j = \varepsilon, 0, 1, \ldots, a_2, \;\; k = \varepsilon, 0, 1, \ldots, a_3 \right\} \tag{7.29}$$

To describe binding to the scaffold, we would also add to the set R the following reactions

$$R' = \left\{ S_{p_1, \cdots, p_i = \varepsilon, \cdots, p_n} + K_i^j \leftrightarrow S_{p_1, \cdots, p_i = j, \cdots, p_n} \right\} \tag{7.30}$$

where the indices run over all values in the range

$$p_i = \begin{cases} \varepsilon, 0, 1, \ldots, a_i, & i \neq j \\ 0, 1, \ldots, a_i, & i = j \end{cases} \tag{7.31}$$

For the three-member scaffold this would be

$$\begin{aligned} R' = &\left\{ S_{\varepsilon jk} + K_1^i \leftrightarrow S_{ijk}, i = 0, \ldots, a_1, \;\; j = \varepsilon, 0, \ldots, a_2, \;\; k = \varepsilon, 0, \ldots, a_3 \right\} \\ &\bigcup \left\{ S_{i\varepsilon k} + K_2^j \leftrightarrow S_{ijk}, i = \varepsilon, 0, \ldots, a_1, \;\; j = 0, \ldots, a_2, \;\; k = \varepsilon, 0, \ldots, a_3 \right\} \\ &\bigcup \left\{ S_{ij\varepsilon} + K_3^k \leftrightarrow S_{ijk}, i = \varepsilon, 0, \ldots, a_1, \;\; j = \varepsilon, 0, \ldots, a_2, \;\; k = 0, \ldots, a_3 \right\} \end{aligned} \tag{7.32}$$

Finally, we have phosphorylation in the scaffold. This can be done either by a protein that is not bound to the scaffold, *e.g.*, for the input signal,

$$R'' = \left\{ S_{p_1, \cdots, p_{i-1} = j < a_{i-1}, p_i = a_i, \cdots, p_n} + K \Leftrightarrow S_{p_1, \cdots, p_{i-1} = j+1, p_i = a_i, \cdots, p_n} \right\} \tag{7.33}$$

where the two-sided double arrow ($\Leftrightarrow$) is used as shorthand for the (possibly bi-directional) enzymatic reaction, or by one that is bound to the scaffold,

$$R''' = \left\{ S_{p_1, \cdots, p_{i-1} = j < a_{i-1}, p_i = a_i, \cdots, p_n} \rightarrow S_{p_1, \cdots, p_{i-1} = j+1, p_i = a_i, \cdots, p_n} \right\} \tag{7.34}$$

or by some combination of the two, all of which must be added to the reaction list R. For the three-slot scaffold with external signal K_4 that activates K_3, we have

$$
\begin{aligned}
R'' &= \{S_{i,a_2,k} \rightarrow S_{i+1,a_2,k}, i = 0, \ldots, a_1 - 1, k = \varepsilon, 0, \ldots, a_3\} \\
&\bigcup \{S_{i,j,a_3} \rightarrow S_{i,j+1,a_3}, i = \varepsilon, 0, \ldots, a_1, j = \varepsilon, 0, \ldots, a_2 - 1\}
\end{aligned}
\tag{7.35}
$$

and

$$
R''' = \left\{S_{ijk} \Leftrightarrow^{K_4}_{Ph_3} S_{i,j,k+1}, \quad i = \varepsilon, 0, \ldots, a_1, \; j = \varepsilon, 0, \ldots, a_2, \; k = 0, \ldots, a_3 - 1\right\}
\tag{7.36}
$$

Typical a_i values for this type of cascade are $a_1 = a_2 = 2$ and $a_3 = 1$.

As an example, let us continue with the above-mentioned three-member cascade that is initiated with K_4. In what follows, we refer to Cellerator, a Mathematica® package that implements the above algorithms. In Cellerator we have defined the function

genReacts[*kinase-name*, *n*, $\{a_i\}$, *phosphatase-name***],**

where *kinase-name* and *phosphatase-name* are the names we want to give to the sequences of kinases and phosphatases, respectively, and n and a_i are as before. The following Cellerator command then generates the above set of reactions (7.20),

```
genReacts[K, 3, {2, 2, 1, 1}, kpase]

      K[2,2]                  K[2,2]                  K[3,1]                  K[3,1]                  K[4,1]
{K[1, 0] ⇄ K[1, 1], K[1, 1] ⇄ K[1, 2], K[2, 0] ⇄ K[2, 1], K[2, 1] ⇄ K[2, 2], K[3, 0] ⇄ K[3, 1]}
      kpase[1]                kpase[1]                kpase[2]                kpase[2]                kpase[3]
```

The input is in the first line while the output is the second line. Alternatively, the user could specify the set of reactions explicitly, or copy the output to a later cell to manually add additional reactions. If RAF has been set up as an alias for K_3 then the rate constants are specified by a content-addressable syntax, e.g., as

```
                             RAFK
storeRateConstant[db, RAF ⇄ RAF*, a1, d1, k1, a2, d2, k2];
                             RAFP
```

corresponding to

$$RAF + RAFK \overset{a_1}{\underset{d_1}{\Longleftrightarrow}} RAF - RAFK \overset{k_1}{\Longrightarrow} RAF^* + RAFK \tag{7.37}$$

and

$$RAF^* + RAFP \overset{a_2}{\underset{d_2}{\Longleftrightarrow}} RAF^* - RAFP \overset{k_2}{\Longrightarrow} RAF + RAFP \tag{7.38}$$

and so forth, where the numbers over the arrows indicate the rate constants (and not enzymes, as with the double arrow notation). Cellerator first translates the five high-order reactions (equation (7.20)) into the corresponding set of 30 low-level reactions. Each low-level reaction (such as intermediate compound formation) is determined by applying the appropriate enzyme-kinetics description, and has a unique rate constant. The low-level reactions are subsequently translated into the appropriate set of 21 differential equations for the eight kinases, three phosphatases and ten intermediate compounds. When scaffold proteins are included (discussed below) these numbers increase to 139 high level reactions, 348 low-level reactions (300 without kinases), and 101 differential equations (85 without kinases).

MAPK PATHWAY WITH SCAFFOLDS: EXPERIMENTAL BACKGROUND

The mitogen-activated protein kinase (MAPK) cascades (Figure 7.1) are a conserved feature of a variety of receptor mediated signal transduction pathways (Garrington and Johnson, 1999; Widmann et al., 1999; Gustin et al., 1998). In humans they have been implicated in transduction of signals from growth factor, insulin and cytokine receptors, T cell receptor, heterotrimeric G proteins and in response to various kinds of stress (Garrington and Johnson, 1999; Putz et al., 1999; Sternberg and Alberola-Ila, 1998; Crabtree and Clipstone, 1994; Kyriakis, 1999). A MAPK cascade consists of three sequentially acting kinases. The last member of the cascade, MAPK is activated by dual phosphorylation at tyrosine and threonine residues by the second member of the cascade: MAPKK. MAPKK is activated by phosphorylation at threonine and serine by the first member of the cascade: MAPKKK. Activation of MAPKKK apparently proceeds through different mechanisms in different systems. For instance, MAPKKK Raf-1 is thought to be activated by translocation to the cell membrane, where it is phosphorylated by an unknown kinase. All the reactions in the cascade occur in the cytosol with the activated MAPK translocating to the nucleus, where it may activate a battery of transcription factors by phosphorylation.
MAPK cascades have been implicated in a variety of intercellular processes including regulation of the cell cycle, apoptosis, cell growth and

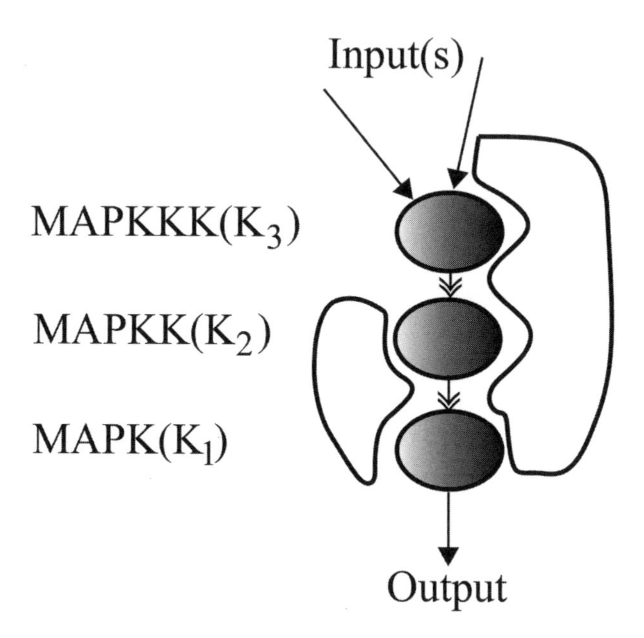

Figure 7.1 The topology of MAPK signaling cascade. Each double arrow represents activation through dual phosphorylation. Two and three-member scaffolds have been identified experimentally and are depicted here.

responses to stress. These molecules are of crucial importance in the development of memory and wound healing. Abnormal changes in MAPK pathway regulation often mediate various pathologies, most notably cancer. This central role of MAPK mediated signal transduction in most regulatory processes makes it an especially attractive research and modeling object.

Signal transduction through a MAPK cascade can be very inefficient unless additional regulatory proteins, called scaffolds, are present in the cytosol. Scaffold proteins nucleate signaling by binding two or more MAP kinases into a single multi-molecular complex. It has been reported previously that scaffolds can both increase and decrease the efficiency of signaling in a concentration dependent manner (Levchenko et al., 2000). In addition they can reduce the non-linear activation characteristics of the cascade. These properties may be crucial for global and local activation of MAPK as scaffold proteins may selectively translocate to small subcellular compartments, thus locally facilitating or inhibiting MAPK activation. In this report we show how the use of Cellerator software package has allowed us to substantially improve our earlier model and study its parametric dependence in a manner not investigated in the preceding report.

As described above, addition of scaffold proteins into the MAPK reaction system results in markedly increased number of states and equations describing transitions between them. Here the benefits provided by Cellerator can really be appreciated, as a simple sequence of commands can lead to automatic description of all reactions involving scaffold-kinase complexes (see Figure 7.2).

In our simulations the first goal was to verify the automatic model generation for scaffold-medicated MAPK cascade as implemented in Cellerator. As a basis for the comparison we referred to our previous report describing a quantitative model of the effect scaffold proteins can play in MAPK mediated signal transduction. When all the assumptions of that model were made again exactly the same solution for the three-member scaffold case was obtained. This convergence of results verified the model generated by Cellerator. In addition, the difficulty of manual generation of all the necessary equations, a limiting factor of the previous study, has now been removed. We thus attempted to study a more detailed model, in which some of the previous assumptions were relaxed.

```
phosphorylationReactions - genScafPhosReacts[S, {2, 2, 1, 1}, K]

{S[0, 2, -1] → S[1, 2, -1], S[0, 2, 0] → S[1, 2, 0], S[0, 2, 1] → S[1, 2, 1], S[1, 2, -1] → S[2, 2, -1],
 S[1, 2, 0] → S[2, 2, 0], S[1, 2, 1] → S[2, 2, 1], S[-1, 0, 1] → S[-1, 1, 1], S[-1, 1, 1] → S[-1, 2, 1],
 S[0, 0, 1] → S[0, 1, 1], S[0, 1, 1] → S[0, 2, 1], S[1, 0, 1] → S[1, 1, 1], S[1, 1, 1] → S[1, 2, 1],
                                                                   K[4,1]
 S[2, 0, 1] → S[2, 1, 1], S[2, 1, 1] → S[2, 2, 1], S[-1, -1, 0] ⇌ S[-1, -1, 1],
             K[4,1]                    K[4,1]                    K[4,1]
 S[-1, 0, 0] ⇌ S[-1, 0, 1], S[-1, 1, 0] ⇌ S[-1, 1, 1], S[-1, 2, 0] ⇌ S[-1, 2, 1],
             K[4,1]                  K[4,1]                  K[4,1]                  K[4,1]
 S[0, -1, 0] ⇌ S[0, -1, 1], S[0, 0, 0] ⇌ S[0, 0, 1], S[0, 1, 0] ⇌ S[0, 1, 1], S[0, 2, 0] ⇌ S[0, 2, 1],
             K[4,1]                  K[4,1]                  K[4,1]                  K[4,1]
 S[1, -1, 0] ⇌ S[1, -1, 1], S[1, 0, 0] ⇌ S[1, 0, 1], S[1, 1, 0] ⇌ S[1, 1, 1], S[1, 2, 0] ⇌ S[1, 2, 1],
             K[4,1]                  K[4,1]                  K[4,1]                  K[4,1]
 S[2, -1, 0] ⇌ S[2, -1, 1], S[2, 0, 0] ⇌ S[2, 0, 1], S[2, 1, 0] ⇌ S[2, 1, 1], S[2, 2, 0] ⇌ S[2, 2, 1]}
```

Figure 7.2 The implementation of automatic generation of the MAP kinases activation reactions (through phosphorylation) in the scaffold in the Cellerator environment. All the possible scaffold states (species) are generated as are the transition reactions between them. The indexes in the parentheses indicate the phosphorylation status of the kinase in the corresponding position, with –1 corresponding to the absence of the kinase from the scaffold complex. K[4,1] represents the external kinase activating the first MAP kinase (MAPKKK) in the cascade.

The use of Cellerator has allowed us to perform systematic sensitivity analyses of the assumptions made in our description of the role of scaffold proteins in MAPK cascade regulation (Levchenko et al., 2000). We previously described dual MAPKK and MAPK phosphorylation within the scaffold to proceed as a single step (processive activation). This is substantially different from a two-step dual phosphorylation sequence occurring in solution. In this distributive activation, the first phosphorylation event is first followed by complete dissociation from the activating kinase and subsequently the second phosphorylation reaction occurs. The assumption of processive phosphorylation in the scaffold has some experimental basis. Mathematically, it is equivalent to assuming that the rate of the second phosphorylation reaction is fast compared to the first reaction. Although this assumption was partially relaxed in our previous report, no systematic study of relaxation of this assumption has been performed. Using Cellerator we performed a systematic investigation of the role of increasing or decreasing the rate of the second phosphorylation within the scaffold compared to reactions in solution. The results for the case when the two rates are equal are presented in Figure 7.3. It is clear that relaxation of this assumption results in a substantial decrease of efficiency of signal propagation.

Similar simulations were performed to investigate the effect of allowing formation of a complex between MAPKKK in the scaffold and MAPKKK-activating kinase, as well as the effect of allowing phosphatases to dephosphorylate scaffold-bound kinases. In all cases the parameter values used in simulation are equal to those used for corresponding reactions in solution (for the full list of parameters see Levchenko et al. (2000)). The results are presented in Figure 7.3. Again, new assumptions resulted in substantial down-regulation of efficiency of signal propagation. It is of interest that the position of the optimum scaffold concentration (at which the maximum signaling is achieved) is insensitive to making these new assumptions. This agrees with the analysis in (Levchenko et al., 2000), which suggested that the position of the optimum is determined only by the total concentrations of the kinases and their mutual interaction with the scaffold.

DISCUSSION AND FUTURE DIRECTIONS

We have shown that automatic model generation can simplify the transition from an informal, cartoon-based description of a reaction pathway (or a network of pathways) to a system of differential equations. This transition is obtained via a rigorous description of enzymatic kinetics and other biochemical processes and is implemented utilizing symbolic translation. In addition to facilitating the potentially burdensome task of correctly writing out all of the necessary equations, this methodology provides an explicit and flexible way of controlling successive stages of model creation. Furthermore, user intervention is possible both at the stage of conversion of an informal pathway description into a set of chemical reactions and at the later stage of mapping these reactions to the correspond-

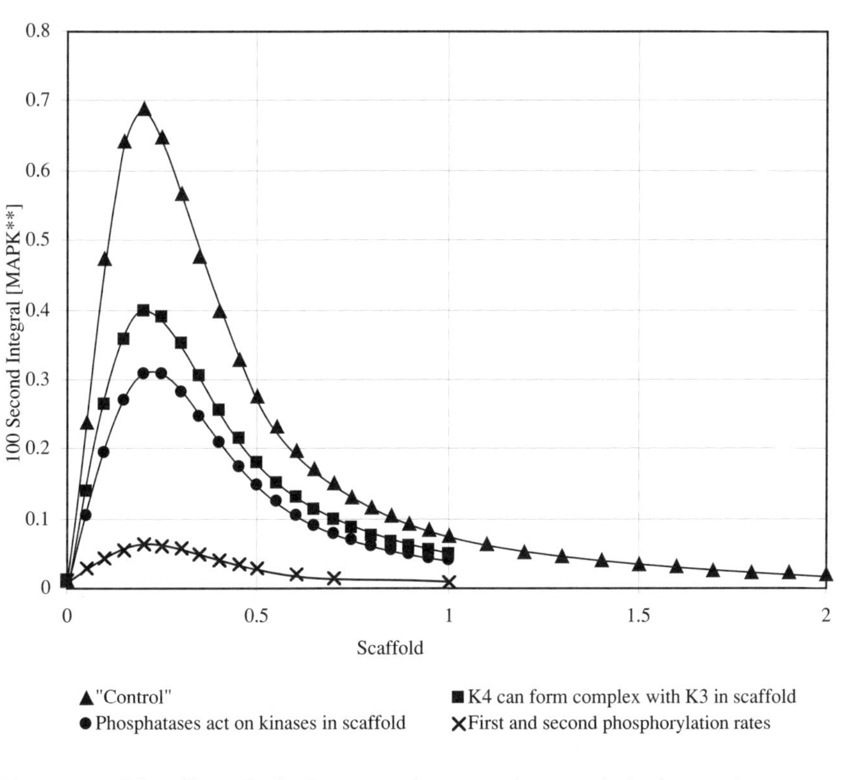

Figure 7.3 The effect of relaxing several assumptions made in the previous report. The time integral of free dually phosphorylated MAPK over first 100 sec is plotted vs. scaffold concentration. The "control" curve reproduces the data with all the assumptions made previously, whereas the other curves represent the results of relaxation of these assumptions as described in the legend. All data are obtained using the Cellerator package and are plotted in Microsoft Excel.

ing mathematical forms. This flexibility is likely to increase the ability of the user to participate in building and modifying the model at a level limited only by his or her expertise.

We have demonstrated the automatic generation of symbolic differential equations using a generic three-member scaffold, the MAPK cascade mediated signaling system. The implementation that we have presented – Cellerator – is capable of generating and solving these 101 differential equations, a task not achieved in the previous detailed study of the effect of scaffolds. Such automated model generation will prove especially useful in describing even more complex biochemical reactions that involve the formation of multi-molecular complexes. Such complexes may exist in numerous states, each requiring a corresponding equation for its dynamical description. Because of the combinatoric expansion of reaction possibilities, correctly writing out all of these equations by hand rapidly becomes impossible.

We intend to pursue the research into role of scaffolds in signal transduction regulation using this new tool. In particular we intend to use extended indexing to specify reactions occurring in various sub-cellular compartments. This will facilitate the study of the effect of scaffold translocation to the cell membrane observed in gradient sensing and other important regulatory processed. In addition we will attempt to develop our algorithm to allow for scaffold dimerization, an experimentally observed phenomenon.

Currently, Cellerator is "tailor-made" for modeling events in a linear pathway mediated by sequential covalent modification. It is within our immediate plans to make the code more universal to include other canonical forms and variable structure systems. In particular, we are in the process of adapting Cellerator to two test cases: NF-κB and PKA pathways. Consideration of these pathways will necessitate implementation of the elementary reactions describing transcription, translation and protein degradation. In addition, complex formation will be considered as a high level reaction leading to an activation step within the pathway.

ACKNOWLEDGEMENTS

We have benefited from discussions with B. Wold and H. Bolouri. This work was supported in part by the Whittier Foundation, the Office of Naval Research under contract N00014-97-1-0422, the ERATO Kitano Symbiotic Systems Project, the NASA Advanced Concepts program and by a Borroughs-Wellcome Fund Computational Molecular Biology Postdoctoral Fellowship to A.L.

References

Crabtree, G.R. and Clipstone, N.A. (1994). Signal transmission between the plasma membrane and nucleus of T lymphocytes. *Annu. Rev. Biochem.* 63:1045–1083.

Garrington, T.P. and Johnson, G.L. (1999). Organization and regulation of mitogen-activated protein kinase signaling pathways. *Curr. Opin. Cell. Biol.* 11:211–218.

Gustin, M.C., Albertyn, J., Alexander, M., and Davenport, K. (1998). MAP Kinase Pathways in the Yeast Saccharomyces cerevisiae. *Microbiol. Mol. Biol. Rev.* 62:1264–1300.

Kyriakis, J.M. (1999). Making the connection: coupling of stress-activated ERK/MAPK (extracellular-signal-regulated kinase/mitogen-activated protein kinase) core signaling modules to extracellular stimuli and biological responses. *Biochem. Soc. Symp.* 64:29–48.

Levchenko, A., Bruck, J., Sternberg, P.W. (2000). Scaffold proteins may biphasically affect the levels of mitogen-activated protein kinase signaling and reduce its threshold properties. *Proc. Natl. Acad. Sci. USA* 97(11):5818–5823.

Putz, T., Culig, Z., Eder, I.E., Nessler-Menardi, C., Bartsch, G., Grunicke, H., Uberall, F., and Klocker, H. (1999). Epidermal growth factor (EGF) receptor blockade inhibits the action of EGF, insulin-like growth factor I, and a protein kinase A activator on the mitogen-activated protein kinase pathway in prostate cancer cell lines. *Cancer Res.* 59:227–233.

Sternberg, P.W. and Alberola-Ila, J. (1998). Conspiracy theory: RAS and RAF do not act alone. *Cell* 95:447–450.

Widmann, C., Gibson, S., Jarpe, M.B., and Johnson, G.L. (1999). Mitogen-activated protein kinase: conservation of a three-kinase module from yeast to human. *Physiol. Rev.* 79:143–180.

8 Modeling Large Biological Systems From Functional Genomic Data: Parameter Estimation

Pedro Mendes

A very positive outcome of the emerging"omic" disciplines in biology (genomics, proteomics, etc.) is the trend away from extreme reductionism to systems descriptions and analyses. Data is now becoming available consisting of simultaneous measurements of thousands of cellular components such as mRNA and proteins. If sequences of these are put together they will form movies of the cellular machinery in action, and it should be possible to build dynamic models that describe it. Such comprehensive models will represent explicitly large numbers of biochemical reactions at some level of detail. A challenging step due to the non-linearity of these models in constructing and tuning them is to estimate the large set of numerical parameters that they contain. The inverse problem of estimating parameter values from variables requires software that combines simulation with a breadth of optimization algorithms capable of searching global minima in large dimensional spaces. Numerical experiments carried out with the program Gepasi are presented, which suggest a combination of evolutionary algorithms and local minimization methods appear to be suitable for this purpose.

INTRODUCTION

Modern experimental biology is moving away from analyses of single elements (genes, enzymes, etc.) to whole-organism measurements. In particular, technologies such as DNA microarrays (Shalon et al., 1996) and chips (Lockhart et al., 1996) and various mass spectrometry methods (Pennington et al., 1997; Ducret et al., 1998) provide the means to measure the levels of thousands of mRNA and protein species in a single experiment. Technologies for measuring large sets of small metabolites are also being developed (Oliver et al., 1998; Trethewey et al., 1999). These approaches of measuring cellular components at the genomic scale are sometimes referred to as *transcriptomics*, *proteomics* and *metabolomics*. What makes these different from previous molecular biology approaches is that by nature of measuring so many components of the cell, one does not have to decide *a priori* which ones are more likely to be affected by some specific experimental manipulation of the system. Instead, we now have the luxury of making unbiased observations. By measuring a large proportion of the cell's components one obtains a near-complete snapshot of the average state of cell culture or tissue. These technologies have mostly been discussed in the context of aiding in identifying the function of some 40% of all known open reading frames that are still unidentified - the functional genomics agenda. Usually "function" is taken to mean the name of the protein corresponding to the gene or the nature of the immediate molecular action of that protein (Oliver, 1996; Kell and King, 2000), for example that a certain gene codes for a protein phosphatase. This molecular function, though, is ultimately only a piece of the cellular machinery and to understand and predict how cells behave we better look at how the various cellular components work together in the whole system - the systems biology agenda.

It is important to mention that the systems biology agenda is itself not new and that much work has been dedicated to it in the past, mostly from theoretical angles (Bertalanffy, 1964; Rosen, 1970; Kacser and Burns, 1973; Savageau, 1976) but also computationally (Garfinkel et al., 1961; Higgins, 1965). However it is only now, with the advances in transcriptomics, proteomics, and metabolomics, that we are starting to be in a position to obtain data in the amount and extent that is required to decode the complex behavior of whole living cells (from a systems perspective). It is fortunate that indeed much theory and numerical methods are already in place to allow us to make good use of these rich data sets. That is the subject of this manuscript.

These novel "omic" technologies are now being used by a growing number of laboratories producing large data sets that contain detailed characterizations of the state of cells. The progression of publications presenting such results suggests that soon it will become laboratory routine to capture snapshots of nearly all the mRNAs, proteins and metabolites present in any specific cell type. Biological research is undergoing a transformation from manual (hand-craft) measurement of single cellular components towards an automatic (high-throughput) parallel measurement of large numbers of components. Given the large numbers of components and the complexity of their interconnections it is no longer productive or even practical to interpret our observations based on the reductionist approach that was dominant in molecular biology. We will be forced to replace these single-molecule qualitative models of cell function by others that consider the cell as a whole, and which can explain quantitatively its emergent properties. In the sections that follow I will discuss the issues, biological, mathematical and computational, surrounding the challenge of constructing dynamical models of cells from those rich data sets. In the section *Simulation of biochemical kinetics* I will present the type of predictive mathematical models that can be used to "play the movie". In *Parameter estimation through optimization*, numerical optimization methods are presented which can be used to automatically tune a model to experimental data. *Reverse engineering a simple system* presents a numerical experiment illustrating how one can reverse engineer biological systems and discusses which optimization methods are most likely to perform satisfactorily with very large data sets. Finally, a discussion of the results puts in perspective the issues that appear to be hardest to overcome and suggests that rather than being deluged with data, biologists will soon be hungry for much more.

SIMULATION OF BIOCHEMICAL KINETICS

We are interested in the dynamics of biochemical systems at the level of pools (or populations) of molecular species. A pool is a set of molecules that cannot be distinguished from each other. These pools are the components of the system, the biochemical network of a cell type, and are thus the variables of the mathematical models required for simulation. The network is formed of biochemical reactions and transport steps that connect several pools. The dynamics (or perhaps more appropriately the kinetics) of each pool can be described mathematically by an ordinary differential equation (ODE) such as Equation 8.1, the system being the collection of all equations representing all the pools.

$$\frac{dx_i}{dt} = \sum f(x) - \sum g(x) \tag{8.1}$$

Equation 8.1 has a convenient form in which all positive components, $f(x)$, correspond to the steps leading to this pool, and the negative components, $g(x)$, to the steps leading out of the pool (given that in general reactions are reversible it is necessary to first establish a positive direction of flux). This allows for automation of the process of translating the network into differential equations, which is conveniently implemented in the simulator Gepasi (Mendes, 1993, 1997). It is also this special form that allowed the development of algorithms that resolve the structural properties of the system (those that arise solely from the connection scheme) (Reder, 1988; Schuster et al., 1999).

The mathematical representation of the system in the form of equations like 8.1 describes the evolution of the system in time. Thus, given a certain initial state of the system, those equations can be used to describe the changes in all pool sizes (concentrations). They also allow one to calculate steady states of the system, which are zeroes of the system of equations. For the time evolution, one typically uses an ordinary differential equation solver, such as LSODA (Petzold, 1983), for the steady state a combination of the Newton-Raphson method with an ODE solver is an efficient strategy (Mendes, 1993). Several software packages exist (Sauro,

1993; Mendes, 1993; Ehlde and Zacchi, 1995; Schaff et al., 1997; Tomita et al., 1999) that allow one to construct mathematical models of biochemical networks and simulate their behavior. These are essential tools for systems biology.

The single largest benefit of computer simulation of biochemical networks is that it allows the exploration of the behavior of the model very easily. Simulation allows one to easily ask "what-if?" type of questions, and with the speed of today's computers thousands of these questions can be asked in short time. Rather than a substitute for experiments, such simulations should be seen as excellent means of generating hypotheses about the real system, themselves they provide no real answers. It is then with experiments that we can investigate if the model was correct. If not, then it is time to refine it such that it can represent the newfound evidence and we go back in this process. Simulation is an essential tool for modeling and should always be part of the scientific process. (Of course, even when we do not construct mathematical computer models, we still do simulations in our own thoughts, they are just less rigorous than if they were made explicit.)

Models are intentionally smaller representations of real entities. The success of modeling is exactly the ability to ignore those features of the real system that are irrelevant to the phenomenon studied. So models are by definition simpler than the real system but not in any random way: the model must still capture the essential features of the phenomenon. The art of modeling resides in the process of removing the irrelevant features while retaining the essential ones. Perhaps one could now think that these modern whole-cell technologies are not so useful after all. It may seem at first counterproductive that we should use a technique that floods us with information about all (or at least a large number) of the system's components when we are only interested in a restricted part. Why not just measure those components that we think are important? The answer is simply that at this stage we do not know which ones are important and which are irrelevant — such measurements actually provide a means for us to decide, knowing that very little is escaping our eyes. Unbiased observation is essential for successful model construction!

Traditionally, computer models of biochemical networks only represent the reactions that form the carbon flow of the small molecules, the metabolic pathway. This assumes that the concentrations of the proteins do not change which then implies that the concentrations of mRNA and ribosomes are also constant. In these conditions, the genetic component can be regarded as invariant and so a parameter of the model. Equally, one can also model the genetic subsystem on its own, without considering the metabolic part. The assumption now is that metabolism is so fast that it is essentially in equilibrium and its state is dictated by the genetic part. In both cases, one has to postulate that the genetic and metabolic parts are decoupled. While there are conditions such that these two subsystems may be decoupled (due to very different timescales (Bentem et al., 2000)), it is easy to argue that this is not always the case and may even not be that common. In the same spirit of not deciding *a priori* something that can be observed in the data anyway, perhaps all kinetic models for genomics should include both metabolism and gene expression.

Figure 8.1 shows how one would represent a single metabolic reaction converting metabolite A to B including the genetic component. The enzyme E and its corresponding mRNA are represented explicitly as are the steps that synthesize and degrade them. The synthesis and degradation of the macromolecules if modeled as single steps represent an aggregation of various molecular processes, such as initiation, elongation, termination and splicing for the mRNA synthesis. One can make the model more detailed than Figure 8.1, but at some stage, it becomes counterproductive to do so if indeed it is the systemic properties that we aim to understand. Given that these four steps of synthesis and degradation of E and mRNA are phenomenological rather than based on molecular mechanisms, it is important to examine a little how their rates should be expressed mathematically. Table 8.1 shows proposals for the kinetic functions associated with these steps. Important features to note are: *i)* the rate of mRNA synthesis is constant only in the absence of inhibitor and activator, these can act cooperatively but cannot make the rate go negative; *ii)* the synthesis of protein is a saturable function on the concentration of mRNA; *iii)* these functions ignore the concentration of precursors (nucleotides and amino acids), therefore these are assumed constant. The functions proposed here are just indications of how one could build such kinetic functions, obvi-

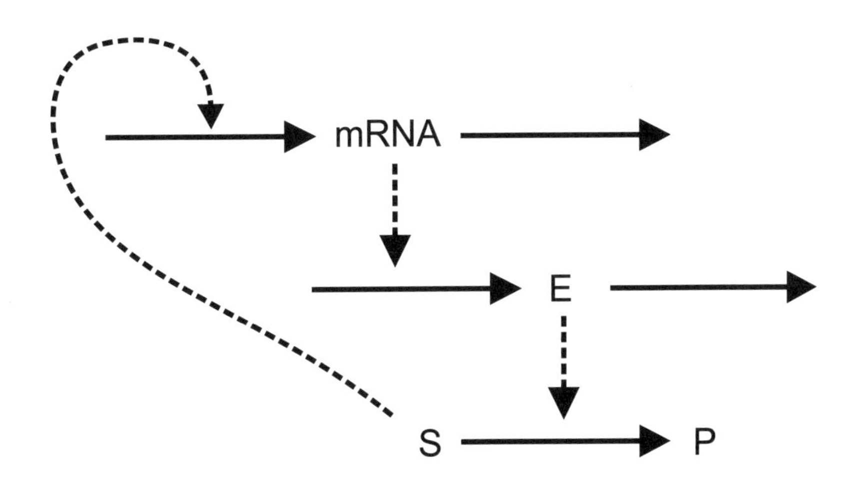

Figure 8.1 Model for a single metabolic reaction including the genetic component. Solid arrows represent mass transfer, dashed arrows represent kinetic effects. mRNA activates production of E which in turn activates the metabolic reaction. The substrate S is here an activator of the transcription step.

Table 8.1 Phenomenological rate functions for synthesis and degradation of mRNA and protein.

Step	Rate	Parameters	Variables
mRNA synthesis	$\frac{V}{\left(1+\frac{I}{K_i}\right)^{n_i}\left(1+\frac{K_a}{A}\right)^{n_a}}$	V, K_i, K_a, n_i, n_a	I, A
mRNA degradation	$k_g G$	k_g	G
protein synthesis	$\frac{VG}{K_{0.5}+G}$	$V, K_{0.5}$	G
protein degradation	$k_p P$	k_p	P

NOTE: V is a basal rate; k_g and k_p are first order rate constants; K_i, K_a and $K_{0.5}$ are affinity constants; n_i and n_a are degrees of cooperativity

NOTE: G is the concentration of mRNA, P is the concentration of protein, I is the concentration of an inhibitor and A the concentration of an activator

ously there will be circumstances in which one needs to add further details to these functions in order to better explain the data. As an example consider the case of modeling cells which live in a medium limited in amino acid sources, then the rates of protein synthesis should include the dependence on that (or those) limited amino acids. The equation for the mRNA synthesis can be easily modified to include more activators or inhibitors, by multiplying additional terms of the same form (this assumes they act independent of each other).

With these tools one should be able to construct a model of the known biochemistry of a specific cell, including the hierarchical levels of metabolism, protein synthesis and gene expression. Such a model would be considerably large. Micro-organisms such as the yeast *Saccharomyces cerevisiae* have around 6000 genes, therefore 6000 mRNA species and 6000 proteins. Actually, the mRNA and protein species would be less than the number of genes as some code for tRNA and several proteins are composed by more than one type of polypeptide. This approximation is enough for this purpose, though. The number of small metabolites is somewhat less (due to the high interconnectedness of the metabolic map), perhaps one order of magnitude. In total this would mean some 12600 variables with 30000 different reactions in the model. By all measures such a model has to be considered very large and will require special attention since the scale implies several problems that are not usually associated with smaller models. Until now, published biochemical kinetic models have usually well under 100 variables and reactions.

So far this description has concentrated on how one simulates the behavior of biochemical networks through the construction and solution by computer of mathematical models. But how does one construct such models from observations of the system? This consists of three phases:

1. identification of relationships between the various model variables (the biochemical reactions, their substrates, products and effectors)
2. identification of mathematical functions that appropriately express the

rate of each reaction as a function of the substrates, products and effectors

3. estimation of the numerical values of the parameters that are included in the mathematical functions above

Steps 1 and 2 fall outside the scope of this text. It is perhaps relevant to say here that they are far from being solved yet, especially step 2. Nevertheless, below it will be assumed that one can carry out these two steps by some means and the rest of this work will be focused on the last step, the estimation of the parameters of the model. Once that has happened, running the computer simulation is like playing a movie that has been constructed from the single snapshots (experimental data). Furthermore, we can use the same model to hypothesize about how the biological system would respond to other types of changes in its environment. That is the power of computer simulation of biochemical networks.

PARAMETER ESTIMATION THROUGH OPTIMIZATION

In the real biological system, the metabolite, protein and mRNA concentrations are controlled by a number of parameters such as time, physicochemical properties and the state of the environment around the system. The state of a cell thus depends on its temperature, on the kinetic constants of its various biochemical processes, and on the concentration of nutrients and other chemical compounds. In simulation, one does indeed start with the parameters (kinetic constants and concentrations of external substrates and products) and uses the computer to calculate the concentrations of the metabolites, proteins and mRNAs along time. In contrast to this, what one does in an experiment is to measure the metabolites, proteins and/or mRNA aiming to determine the values of the parameters. This is known as an inverse problem (Mendes and Kell, 1996) and it is similar to what happens in other areas of science and engineering such as inverse kinematics (what forces to apply in a mechanical device to make it reach a certain point in space), and crystallography (given a X-ray diffraction pattern how were the atoms arranged in the crystal).

This inverse problem of estimating the parameters of a large biochemical network from observations of concentrations is in essence not new. In enzymology the exact same problem has been routinely dealt with for a long time, where one uses time courses or initial rates of reaction to determine kinetic parameters of a single enzymatic reaction. The major difference to what is proposed here is the scale, whereas in whole-cell models we have not one but a large number of simultaneous biochemical reactions, thus a very high number of parameters to estimate.

Irrespective of scale and whether one uses an automatic method or prefers to do it manually, parameter estimation is done according to this procedure (see also Figure 8.2):

1. compare the experimental values with simulated ones

2. if the difference is small enough, stop, otherwise adjust parameters in the model
3. simulate the experiment
4. go back to 1

Step 2 is indeed where all the action occurs, more specifically in the process of adjusting the parameters. In enzyme kinetics, step 2 can be carried out using linear regression if the rate equation can be linearized in the parameters, such as in the double-reciprocal plot (Lineweaver and Burk, 1934). More recently it has been argued (Duggleby, 1986; Johnson, 1992) that nonlinear regression is more appropriate. Here one uses a numerical optimization method to carry out step 2 above, usually the Levenberg-Marquardt (Levenberg, 1944; Marquardt, 1963) method.

The process of measuring the distance between data and model (step 1 above) requires one to carry out simulations (step 3), which are in a sense the inverse of optimization. In simulation we calculate the values of the variables given the parameters, in optimization we do exactly the opposite (Mendes, 1998). The actual comparison of real and simulated data is usually done through a function consisting of the sum of the squares of the difference between the measured and simulated values:

$$\sum_i (x_i - y_i)^2 \tag{8.2}$$

The algorithm above requires the combination of optimization routines with software that is capable of simulating biochemical networks (Mendes, 1998). The program Gepasi (Mendes, 1993, 1997, 1998) fulfils this requirement as it provides means to carry out simulation and optimization using several different optimization methods. The program is also specialized to internally construct the appropriate sum of squares when it is supplied with a file containing experimental data. This software can be used equally for single-enzyme problems and for whole-cell parameter estimation using functional genomics data. Figure 8.2 depicts the process of carrying out parameter estimation for biochemical dynamics problems. This figure clearly shows how simulation and optimization are somewhat opposites but that both have to be combined to estimate the parameter values.

It is well known (e.g. (Wolpert and Macready, 1997)) that no single numerical optimization algorithm is best for all problems. An optimization method's performance is dictated by the problem and the data themselves, and vary considerably. Fortunately there are many different numerical optimization algorithms, the most used fall in three classes: *i)* gradient search, *ii)* deterministic direct search and *iii)* stochastic direct search.

Gradient search methods look for the nearest minimum of the function by following a direction that is determined from derivatives of the func-

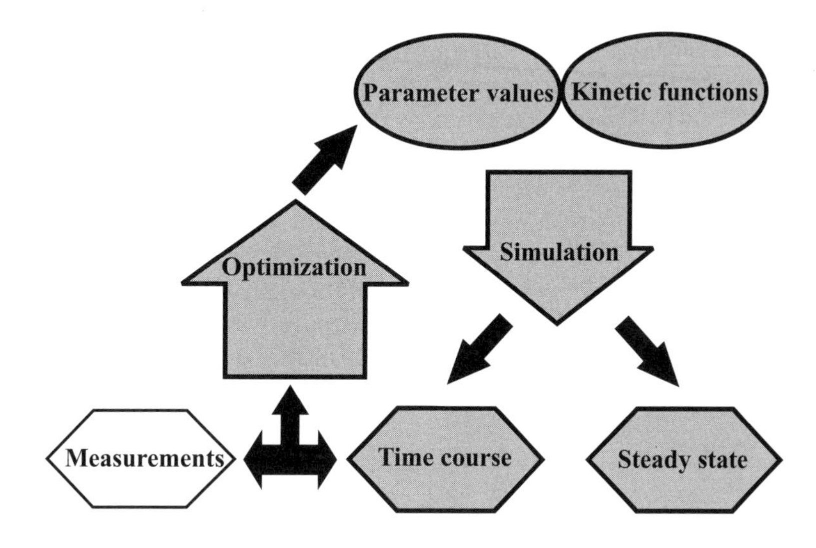

Figure 8.2 The simulation-optimization method for tuning models. Given a set of kinetic functions with numeric values for their parameters simulation calculates a steady state or a time course. The distance between these and (experimental) measured values is computed; optimization methods change the parameter values based on this distance, closing the loop.

tion. Many have originated from the Newton method that uses first and second derivatives, while steepest descent uses only first derivatives. With noiseless data the Newton method would converge to the minimum of a quadratic function in exactly one iteration. There are indeed many methods in this class, the most popular being the Levenberg-Marquardt (Levenberg, 1944; Marquardt, 1963) method and some quasi-Newton methods like L-BFGS-B (Byrd et al., 1995). One common feature to all these methods is that when they converge it is always to the minimum that is closest to the initial guess. This has severe negative implications to our problem because the sum of squares function has several minima and the one closest to the initial guess has low probability of being the global minimum (the best solution).

Direct search methods are those that do not require or calculate derivatives to minimize an objective function. A number of these are deterministic, which means that they do not use random numbers and anytime they run with the same data and initial guess will provide the same answer. These deterministic direct search methods employ strategies which are based on keeping a memory of a number of previous candidate solutions and deciding on how to go down on the error hyper-surface based on that limited memory. Popular methods of this type are those described by Nelder and Mead (Nelder and Mead, 1965) and Hooke and Jeeves (Hooke and Jeeves, 1961). One advantage of these methods is that they are more robust than those based on gradients even though they may converge slower than the latter for well-behaved functions.

Finally, stochastic direct search methods base their operation on random numbers and thus do not in general provide the same answer when run with the same initial guess and data. Methods of this class include evolutionary algorithms (Bäck and Schwefel, 1993) and simulated annealing (Kirkpatrick et al., 1983). The former are based on biological evolution and population dynamics, evolving a population of solutions towards the minimum. Simulated annealing minimizes functions by following a process analogous to that of producing perfect crystals. The major advantage of these methods is that they can find global minima, however this is at the expense of much larger run times that the methods of the other two classes.

Nonlinear least squares in enzyme kinetics has been mostly carried out with gradient descent methods (mainly the Levenberg-Marquardt method) or deterministic direct search methods that are local minimizers. There are arguments pertaining to this being a poor choice due to the sum of squares function often having several minima (Mendes, 1998). The problem is more severe when the number of parameters is very large, such as the case with whole-cell models (or even models with only a few enzymatic reactions). It seems that for these we will have to resort to methods that are able to locate global minima, like stochastic direct search methods. One further problem of these large models is that the data from microarray and proteomics experiments is currently very noisy. There should be little expectation that parameter values should be determined to any great precision. Neither should that be important as long as the behavior of the real system is reproduced by the model. The target should be to obtain models that can be used to extrapolate without much error and that carry very similar properties to those observed in the real system. It is important to remember that it is not the actual parameter values we care for but that the model reproduces the observed phenomena!

REVERSE ENGINEERING A SIMPLE SYSTEM

To demonstrate how the procedure proposed here would work, an example is now described. This example is also the seed for a larger scale study of optimization methods for parameter estimation in whole-cell models, currently in progress in this laboratory. In the example shown here, experimental data will not be used, instead a model of a pathway will be adopted and data will be generated by simulation, which will correspond to the experimental data mentioned in the previous section. This choice is important because one of the present aims is to compare the performance of several algorithms. That can only be possible when the source of the data is known, so that one can compare the estimated parameters with the real ones. Were data from an actual experiment to be used, it would be

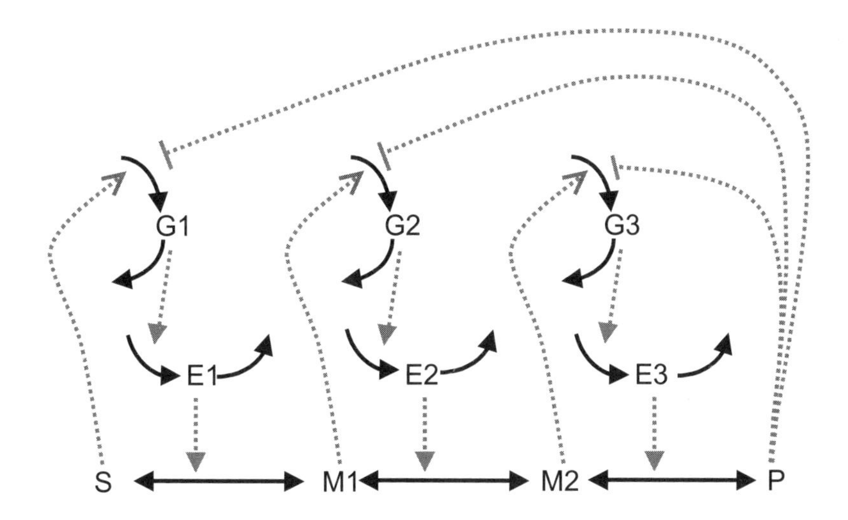

Figure 8.3 A model of 3-enzyme pathway with gene expression. Solid arrows indicate mass transfer reactions and point to the positive direction of flux (but are chemically reversible). Dashed arrows indicate activation and dashed curves with blunt ends indicate inhibition

impossible to judge how good the parameter estimation worked since we would never know the values of the parameters in the real system. It is also useful to be able to control the level of noise in the data as this allows for a study of robustness of the optimization methods to noise. Using this approach it is possibly to add noise with well-defined properties to the data.

We will consider a small pathway of three enzymatic steps including, as discussed above, the enzymes and mRNAs explicitly. Figure 8.3 contains a diagram illustrating the network of reactions and kinetic effects (feedback loops). This is a rather small model compared with those discussed in the previous sections. Indeed, it is not a whole-cell model or even close to that, but later it will become obvious that this is already very hard to reverse engineer from data. A long-term objective is to extend this analysis to models of large biochemical networks, but for now we can learn something out of this exercise.

The model pathway, as in Figure 8.3, was defined in the program Gepasi[1] and then the kinetics for each step selected. The rate of each enzymatic step follows Equation 8.3, where A is the substrate, P is the product, E is the enzyme, K_{eq} is the equilibrium constant of the step, k_{cat}, K_m^a and K_m^p are kinetic constants.

1 available at http://www.gepasi.org

Table 8.2 Kinetic parameter values of 3-step biochemical model.

Step type	Parameter	Value
enymatic reaction	K_{eq}	1
enymatic reaction	k_{cat}	1
enymatic reaction	K_m^a	1
enymatic reaction	K_m^p	1
protein synthesis	V	0.1
protein synthesis	$K_{0.5}$	1
protein degradation	k_p	0.1
mRNA synthesis	V	1
mRNA synthesis	K_i	1
mRNA synthesis	K_a	1
mRNA synthesis	n_i	2
mRNA synthesis	n_a	2
mRNA degradation	k_g	1

NOTE: All steps of the same type have the same numerical parameters

$$\frac{k_{cat}\,E\,\frac{\left(A-\frac{P}{K_{eq}}\right)}{K_m^a}}{1+\frac{A}{K_m^a}+\frac{P}{K_m^p}} \tag{8.3}$$

The rates of the transcription steps follow equation 8.4:

$$\frac{V}{1+\left(\frac{I}{K_i}\right)^{n_i}+\left(\frac{K_a}{A}\right)^{n_a}} \tag{8.4}$$

This is similar to the one presented in Table 8.1, except for the denominator, which has a slightly different form — in this case the action of the inhibitor and activator are not independent. The protein synthesis, mRNA and protein degradation steps follow the exact same kinetics as presented in Table 8.1. The numerical values of the constants are summarized in Table 8.2. To obtain a reference state for this model, the concentration of substrate S was set to 2 and the product P to 0.5. A preliminary simulation was then run to calculate the steady-state concentrations and these were set as the initial state of the system.

Having put the model in a steady state, a perturbation was then added: the concentration of the product of the pathway (P) was decreased to 0.05, which causes the pathway to adjust to a state of higher disequilibrium, therefore a larger flux from S to P. The model was once again simulated, this time to follow how the initial concentrations adjust to the new environmental conditions. Gepasi was instructed to sample the concentrations of M1, M2, E1, E2, E3, G1, G2, and G3 along time. This corresponds to the experimental situation in which one would measure the transcriptome, the proteome and the metabolome for each sample. No noise was added to these data, which is rather unrealistic. This was on purpose, so that a best-case scenario could be established. The effects of noise superimposed to the data will be studied later. The information on the concentrations of those chemical species against time corresponds to the "experimental" data that will be used for parameter estimation. This time course was sampled with 20 time points.

Once in possession of the "experimental" data it is then time to attempt to go back and recover the original parameter values from these data. Most methods require that an initial guess of the parameter values be made. Here this was set to the arbitrary value of 10^{-7}, except for the degrees of cooperativity that were set to unity and the equilibrium constants, which were left at their nominal value. There will be no attempt to estimate the latter, since it is assumed that they can be determined from in vitro experiments. The value 10^{-7} was chosen to be very far from the actual value, since in a real setting one has high probability of guessing a value badly. A total of 36 parameters are then to be estimated. At this stage we will assume that the degrees of cooperativity will lie between 0.1 and 10 and that all other parameters between 10^{-12} and 10^{6}. This means that in each dimension (except those of the degrees of cooperativity) we have 18 orders of magnitude to search, a very hard problem indeed.

Having defined the parameter estimation problem with the simulation software, all that remains to be done now is to choose an optimization method and to run the program. Given the nature of this exercise, several methods were applied. Given their popular status in enzyme kinetic non-linear regression, the gradient descent methods were the first to be tried. Of these, L-BFGS-B was the best, but still only managed to reduce the sum of squares to around 0.6, which is a poor solution. The Levenberg-Marquardt method performed very poorly (1.16602), in fact slightly worse than the steepest descent method (1.16599). The worse being NL2SOL (Dennis et al., 1981) (2.53908). The direct search method of Hooke and Jeeves performed better (0.215) than the gradient methods, however that solution is still poor. The evolutionary programming method, one of the available evolutionary algorithms in Gepasi, did a much better job, reducing the sum of squares to around 0.0051. Visual inspection of plots of real versus simulated data indicates that this solution could already be seen as a reasonable qualitative fit.

Evolutionary algorithms are known to be good at pushing the solution towards the vicinity of the minimum, but rather bad at actually reaching it. But that is is exactly what local optimizers are good at. In light of this, the solution obtained above with evolutionary programming was used as a initial guess for the same set of optimization methods that had failed previously. Hooke and Jeeves managed to take the sum of squares further down to 0.0037, while Levenberg-Marquardt diverged away to 0.76. However, when Levenberg-Marquardt was applied to the solution of Hooke and Jeeves, it then converged down to 0.0022. A final iteration of Hooke and Jeeves took the value down to 0.0020, Figure 8.4 illustrates how the model fits the data. This clearly illustrates the convergence requirement of gradient descent algorithms to be close to the solution.

Inspection of Figure 8.4 reveals that although the fit is not perfect, in particular with the time courses of E2, G2 and G3, it appears to be of very high quality from this graphical representation. Some could argue that even thought the time courses are very similar, that the model could have other properties unacceptably divergent from the real system. Indeed in terms of the distribution of control there are considerable divergences, but steady state concentrations and fluxes are very similar. Finally it should be noted that some of the parameters were estimated very badly as seen in Table 8.3.

DISCUSSION

General systems science was a field in much activity in the period of 1950–1980. This attempted to be a unifying approach to science using concepts from engineering. Those early efforts were not very productive in terms of large-scale models, mostly because the technology to do so was not in place. In fact some would argue that even in the present days of microarrays we do not yet have sufficiently good technology to do so, but it would be hard to contest that, at least, this is now in the horizon. Essentially the same happened with the computational aspects, where the computers those days hardly managed to cope with small models (simulated annealing and evolutionary algorithms can easily require millions of simulations to be ran). Nevertheless those early studies gave us a valuable body of theory, which is indeed very useful today. For example Metabolic Control Analysis (MCA) was developed independently by Kacser and Burns (Kacser and Burns, 1973) and Heinrich and Rapoport (Heinrich and Rapoport, 1974) (but see also (Higgins, 1963)). MCA emphasizes the systemic approach to measuring control and regulation in metabolic pathways and nicely links the properties of single components to the global system's properties. More importantly, MCA reveals some emergent properties of the system, such as the summation theorems (Kacser and Burns, 1973; Westerhoff and Chen, 1984). MCA is now used by a growing number of experimental biochemists (Fell, 1992) and the approach seems also to benefit from genomic technologies (Kell and Mendes, 2000) (and vice versa (Oliver, 1996)). The first wave of research in systems biology was motivated by the then new technology of electronics; the current one is motivated by the perspective of being able to measure a large number of variables in parallel and by the abundant (and maybe underused) computing power.

Here a very small example was studied by computer simulation and optimization that pertained to illustrate the issues surrounding the use of genomics data for construction of large-scale cellular models. This study

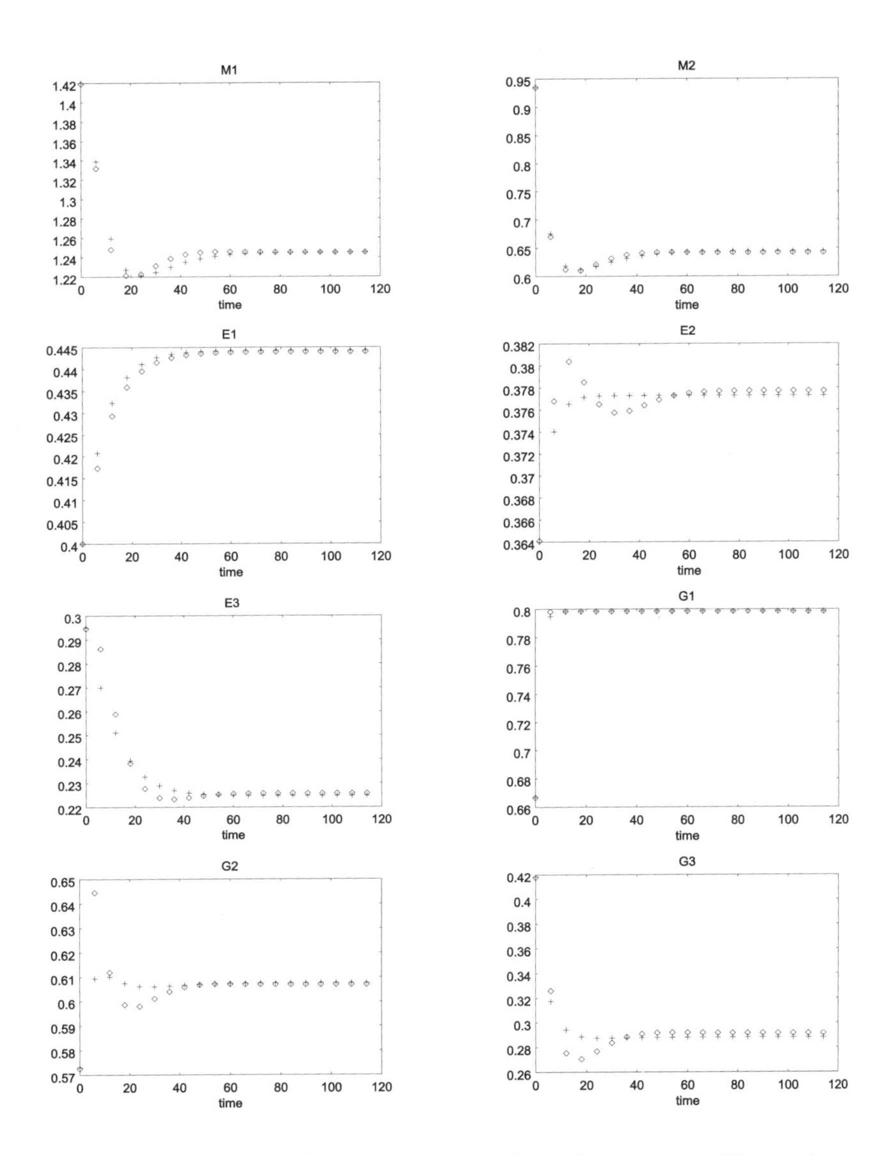

Figure 8.4 Comparison of best regression model and real system. Observed time course plotted with open diamonds, simulated data with crosses.

focused on the numerical optimization algorithms that can be used to fit the model to the data. It was not the intention to discuss the issues involved in the determination of the structure of the pathway from data or the form of the mathematical equations for the reaction rates. The main purpose here was really estimating the numerical values of the parameters contained in those functions.

It is clear from this example that the popular gradient descent methods are inadequate on their own as one cannot guarantee that they will be able to minimize the sum of squares function from any initial guess (they are not robust). Ironically, the popular Levenberg-Marquardt and the highly regarded NL2SOL methods had the worst performances in our example. This is only a reflection of the seriousness of the problem that we will face when instead of dealing with a small 3-enzyme model we will attempt to use gene expression, proteomics and metabolomics data for modeling whole cells. It is by virtue of the high-dimensionality of parameter space that the gradient methods are not robust. These can fail to converge in two distinct situations that depend solely on the data (there are other ways in which they diverge, but those are related to the methods themselves). The first situation is when the derivatives of the sum of squares function of some or all of the parameters are close to zero (a flat region of parameter space). In this case the gradient methods fail to find a descent direction and usually halt prematurely (i.e. as if they had converged to the solution). The second case is when these methods are trapped in a local minimum. Here there is no direction to improve the solution and the methods also halt prematurely. It is clear that solely relying on gradients will be a problem.

Direct search methods, especially those that use random numbers, are robust to local minima and to flat error regions. These methods may take a long time to move but they will eventually do so and therefore are more reliable. Evolutionary algorithms seem especially appropriate for these situations. The method that managed to attract the model towards the solution in the example above was indeed evolutionary programming. Once it got close to the solution, then the gradient descent, and other local optimizers, were able to approximate the solution even better. This suggests that a hybrid algorithm combining evolutionary and gradient methods (e.g. (Pál, 1996)) might be especially well suited to these problems.

For the same problem, evolutionary algorithms usually require considerably longer run times than any of the deterministic methods (gradient or direct search). This can be measured for problems in which all these algorithms converge, such as with an inhibition kinetics data set analyzed by Kuzmic (Kuzmic, 1996) and Mendes (Mendes, 1998). Of course in cases such as that analyzed here, there is no comparison because the deterministic methods did not converge to a good solution. In the example shown here, evolutionary programming took some 3 hours and 1.5 million simulations to run on a Pentium III 600Mhz processor. This number would increase more or less linearly with the number of measured samples and superlinearly with the number of parameters to estimate. Obviously fitting a whole-cell model will require a very long time indeed on a similar machine. Fortunately, evolutionary algorithms are among the simplest to parallelize. Gradient descent methods are not the most sound for this problem since they are harder to code efficiently for parallel computers and don't benefit as much from that as do evolutionary algorithms. There is also evidence that parallel implementations of genetic algorithms not only run faster as produce higher quality solutions (Elketroussi and Fan,

1994). This lends even stronger support for using these methods for large-scale modeling of biochemical systems.

In the example shown here, the best estimation of parameter values was not equally successful for all the parameters (Table 8.3). Some, such as K_m^a and K_m^b (Michaelis constants) of the first two enzymatics steps are indeed estimated to values very far from reality. It could seem a contradiction that an acceptable solution could contain parameters so badly off, but in reality a model will not be equally sensitive to all the parameters. Two explanations are possible for this lack of fit of some model parameters. The first is that if we had better estimates to start with, the optimization would have converged even for these parameters. Indeed there is a good scope of improving the process above by making more educated guesses of the values of the parameters (for example it could be possible to estimate the order of magnitude of the parameters that are rate constants). The second one is that for this time course data, those parameters could change considerably without any observable effect on the time courses — the model would be insensitive to them. In that case one would be advised to simplify the model, somehow removing them.

The second explanation for lack of fit over some parameters is perhaps the correct one here, since the amount of data that was used was very small: only 20 samples of the same time course were used to estimate 36 parameters. Even to estimate standard deviations for the fit would require a minimum of 38 time points in the time course. Although it would have been very easy to collect more data here, this is hardly the case in large-scale experiments. Using the numbers calculated in the introduction, a whole cell model with 30,000 reactions would have on the order of 100,000 parameters, and so would require 100,001 samples to calculate a simple standard deviation! It is best to prepare ourselves to the fact that we are not going to be able to predict all those 100,000 parameters with equal precision. Note that to increase the chance of being able to estimate all parameters one does not have to just sample the same time course with higher frequency, as the data in that time course are highly correlated to each other and so the information content would not increase much (Wahde and Hertz, 2000). It would be necessary to carry out other experiments where the system would be perturbed in different ways. Such comprehensive perturbation experiments are probably not going to happen soon due to limitations on time, money and/or the current technology.

Table 8.3: Parameter values produced by the best estimation.

Step	Parameter	True Value	Estimated Value
S to M1	k_{cat}	1	0.5604
S to M1	K_m^a	1	$3.455 \cdot 10^{-7}$
S to M1	K_m^p	1	$4.467 \cdot 10^{-6}$
M1 to M2	k_{cat}	1	0.4924
M1 to M2	K_m^a	1	$8.413 \cdot 10^{-12}$
M1 to M2	K_m^p	1	$1.225 \cdot 10^{5}$
M2 to P	k_{cat}	1	0.8174
M2 to P	K_m^a	1	0.1299
M2 to P	K_m^p	1	0.01463
E1 synthesis	V	0.1	0.05648
E1 synthesis	$K_{0.5}$	1	0.1015
E1 degradation	k_p	0.1	0.1128
E2 synthesis	V	0.1	0.09027
E2 synthesis	$K_{0.5}$	1	0.09431
E2 degradation	k_p	0.1	0.2356
E3 synthesis	V	0.1	0.03572
E3 synthesis	$K_{0.5}$	1	0.1638
E3 degradation	k_p	0.1	0.1013
G1 synthesis	V	1	0.5466
G1 synthesis	K_i	1	94.11
G1 synthesis	K_a	1	0.07055
G1 synthesis	n_i	2	10
G1 synthesis	n_a	2	0.5431
G1 degradation	k_g	1	0.5887
G2 synthesis	V	1	0.2373
G2 synthesis	K_i	1	2.050
G2 synthesis	K_a	1	0.1861
G2 synthesis	n_i	2	9.973
G2 synthesis	n_a	2	0.6880
G2 degradation	k_g	1	0.3074
G3 synthesis	V	1	0.09607
G3 synthesis	K_i	1	1.685
G3 synthesis	K_a	1	0.01500
G3 synthesis	n_i	2	3.507
G3 synthesis	n_a	2	0.3388
G3 degradation	k_g	1	0.2606

Some say that with all the high-throughput parallel experimental methods available today biologists are already deluged with data. Certainly if we look at the volume of data, reflected for example in the disk space needed to store it, it seems that indeed there is a great deal of information around. However, the calculation in the previous paragraph should enlighten us to the reality that biology needs much more data than it can possibly generate in the near future. In fact the argument should be that we need to collect much more (informative) data than we are doing right now.

References

Bäck, T. and Schwefel, H.-P. (1993). An overview of evolutionary algorithms for parameter optimization. *Evolutionary Computation* 1:1–23.

Byrd, R.H., Lu, P., Nocedal, J., and Zhu, C. (1995). A limited memory algorithm for bound constrained optimisation.*SIAM Journal of Scientific Computing* 16(5):1190–1208.

de la Fuente van Bentem, A., Mendes, P., Westerhoff, H.V., and Snoep, J.L. (2000). Can metabolic control analysis be applied to hierarchical regulated metabolism? MCA versis HCA. In Hofmeyr, J.-H.S., Rohwer, J.H., and Snoep, J.L. editors, *BTK2000: animating the cellular map* pp.191–198. Stellenbosch University Press, Stellenbosch.

Dennis, J.E., Gay, D.M., and Welsch, R.E. (1981). Algorithm 573 — NL2SOL: an adaptive non-linear least-squares algorithm. *ACM Transactions on Mathematical Software* 7(3):369–383.

Ducret, A., van Oostveen, I., Eng, J.K., Yates, J.R., and Aebersold, R. (1998). High throughput protein characterization by automated reverse-phase chromatography electrospray tandem mass spectrometry. *Protein Science* 7:706–719.

Duggleby, R.G. (1986). Progress-curve analysis in enzyme kinetics. Numerical solution of integrated rate equations. *Biochemical Journal* 235(2):613–5.

Ehlde, M. and Zacchi, G. (1995). MIST: a user-friendly metabolic simulator. *Computer Applications in the Biosciences* 11(2):201–207.

Elketroussi, M. and Fan, D.P. (1994) Optimization of simulation models with GADELO: a multi-population genetic algorithm. *International Journal of Bio-medical Computing* 35:61–77.

Fell, D.A. (1992). Metabolic control analysis — a survey of its theoretical and experimental development. *Biochememical Journal* 286:313–330.

Garfinkel, D., Rutledge, J.D., and Higgins, J.J. (1961). Simulation and analysis of biochemical systems. I. representation of chemical kinetics. *Communications of the Association of Computing Machinery* 4:559–562.

Heinrich, R. and Rapoport, T.A. (1974). A linear steady-state treatment of enzymatic chains. General properties, control and effector strength. *European Journal of Biochemistry* 42:89–95.

Higgins, J. (1963). Analysis of sequential reactions. *Annals of the New York Academy of Sciences* 108:305–321.

Higgins, J. (1965). Dynamics and control in cellular reactions. In B. Chance, R.K. Eastbrook, and J.R. Williamson, editors, *Control of energy metabolism.*, pp.13–46. Academic Press, New York.

Hooke, R. and Jeeves, T.A. (1961). "Direct search" solution of numerical and statistical problems. *Journal of the Association for Computing Machinery* 8:212–229.

Johnson, M.L. (1992). Why, when, and how biochemists should use least squares. *Analytical Biochemistry* 206(2):215–25.

Kacser, H. and Burns, J.A. (1973). The control of flux. *Symposium of the Society for Experimental Biology* 27:65–104.

Kell, D.B. and King, R.D. (2000). On the optimization of classes for the assignment of unidentified reading frames in functional genomics programmes: the need for machine learning. *Trends in Biotechnology* 18:93–98.

Kell, D.B. and Mendes, P. (2000). Snapshots of systems — metabolic control analysis and biotechnology in the post-genomic era. In A. Cornish-Bowden and M.L. Cárdenas, editors, *Technological and Medical Implications of Metabolic Control Analysis*, pp.3–25 – see also `http://gepasi.dbs.aber.ac.uk/dbk/mca99bio.htm`. Kluwer Academic Publishers, Dordrecht.

Kirkpatrick, S., Gelatt, Jr, C.D., and Vecchi, M.P. (1983). Optimization by simulated annealing. *Science* 220:671–680.

Kuzmic, P. (1996). DYNAFIT for the analysis of enzyme kinetic data: application to HIV proteinase. *Analytical Biochemistry* 237(2):260–273.

Levenberg, K. (1944). A method for the solution of certain nonlinear problems in least squares. *Quart. Appl. Math.* 2:164–168.

Lineweaver, H. and Burk, D. (1934). The determination of enzyme dissociation constants. *Journal of the American Chemical Society* 56:658–666.

Lockhart, D.J., Dong, H.L., Byrne, M.C., Folliette, M.T., Gallo, M.V., Chee, M.S., Mittmann, M., Wang, C.W., Kobayashi, M., Horton, H., and Brown, E.L. (1996). Expression monitoring by hybridization to high-density oligonucleotide arrays. *Nature Biotechnology* 14:1675–1680.

Marquardt, D.W. (1963). An algorithm for least squares estimation of nonlinear parameters. *SIAM Journal* 11:431–441.

Mendes, P. (1993). GEPASI: a software package for modelling the dynamics, steady states and control of biochemical and other systems. *Computer Applications in the Biosciences* 9(5):563–571.

Mendes, P. (1997). Biochemistry by numbers: simulation of biochemical pathways with Gepasi 3. *Trends in Biochemical Sciences* 22:361–363.

Mendes,P. and Kell, D.B. (1996). On the analysis of the inverse problem of metabolic pathways using artificial neural networks. *BioSystems* 38:15–28.

Mendes,P. and Kell, D.B. (1998). Non-linear optimization of biochemical pathways: applications to metabolic engineering and parameter estimation. *Bioinformatics* 14:869–883.

Nelder, J.A. and Mead, R. (1965). A simplex method for function minimization. *Computer Journal* 7:308–313.

Oliver, S.G., Winson, M.K., Kell, D.B., and Baganz, F. (1998). Systematic functional analysis of the yeast genome. *Trends in Biotechnology* 16(9):373–378. [published erratum appears in Trends Biotechnol 1998 Oct;16(10):447].

Oliver, S.G. (1996). From DNA sequence to biological function. *Nature* 379:597–600.

Pál, K.F. (1996). The ground state energy of the Edwards-Anderson Ising spin glass with a hybrid genetic algorithm. *Physica A* 223:283–292.

Pennington, S.R., Wilkins, M.R., Hochstrasser, D.F., and Dunn, M.J. (1997). Proteome analysis: from protein characterization to biological function. *Trends in Cell Biology* 7:168–173.

Petzold, L. (1983). Automatic selection of methods for solving stiff and nonstiff systems of ordinary differential equations. *SIAM Journal of Scientific and Statistical Computing* 4:136–148.

Reder, C. (1988). Metabolic control theory. A structural approach. *Journal of Theoretical Biology* 135(2):175–201.

Rosen, R. (1970). *Dynamical system theory in biology*. Wiley-Interscience, New York.

Sauro, H.M. (1993). SCAMP: a general-purpose simulator and metabolic control analysis program. *Computer Applications in the Biosciences* 9(4):441–450.

Savageau, M.A. (1976). *Biochemical Systems Analysis.* Addison-Wesley, Reading, MA.

Schaff, J., Fink, C.C., Slepchenko, B., Carson, J.H., and Loew, L.M. (1997). A general computational framework for modeling cellular structure and function. *Biophysical Journal* 73:1135–1146.

Schuster, S., Dandekar, T., and Fell, D.A. (1999). Detection of elementary flux modes in biochemical networks: a promising tool for pathway analysis and metabolic engineering. *Trends in Biotechnology* 17(2):53–60.

Shalon, D., Smith, S.J., and Brown, P.O. (1996). A DNA microarray system for analyzing complex dna samples using two-color fluorescent probe hybridization. *Genome Research* 6:639–645.

Tomita, M., Hashimoto, K., Shimizu, T.S., Matsuzaki, Y., Miyoshi, F., Saito, K., Tanida, S., Yugi, K., Venter, J.C., and Hutchinson, C.A. (1999). E-CELL: software environment for whole-cell simulation. *Bioinformatics* 15(1):72–84.

Trethewey, R.N., Krotzky, A.J., and Willmitzer, L. (1999). Metabolic profiling: a Rosetta Stone for genomics? *Current Opinion in Plant Biology* 2(2):83–85.

von Bertalanffy, L. (1964). Basic concepts in quantitative biology of metabolism. In Kinne, O. and Locker, A., editors, *Quantitative Biology of Metabolism - First International Symposium*, volume 9 of *Helgolander wissenschaftliche Meeresuntersuchungen*, pp.5–37, Helgoland.

Wahde, M. and Hertz, J. (2000). Coarse-grained reverse engineering of genetic regulatory networks. *BioSystems* 55(1-3):129–136.

Westerhoff, H.V. and Chen, Y.-D. (1984). How do enzyme activities control metabolite concentrations? An additional theorem in the theory of metabolic control. *Eur. J. Biochem.* 142:425–430.

Wolpert, D.H. and Macready, W.G. (1997). No free lunch theorems for optimization. *IEEE Transactions on Evolutionary Computation* 1:67–82.

Part IV

Cellular Simulation

9 Towards a Virtual Biological Laboratory

Jörg Stelling, Andreas Kremling, Martin Ginkel,
Katja Bettenbrock and Ernst Dieter Gilles

For a system-level understanding of living cells, a quantitative representation of these systems involving mathematical models and corresponding computer tools is required. Our approach focuses on a modeling concept which relies upon modular structuring of cellular systems focusing strongly on the biomolecular structure of these systems. Mathematical submodels for functional units comprising metabolism and regulation can be aggregated in a hierarchical way to obtain more complex modules. In the Virtual Biological Laboratory, the process modeling tool PROMOT contains an object-oriented knowledge base with reusable modeling entities and enables a purely symbolical model development process via a graphical user interface. The simulation environment DIVA then uses the model library for dynamic simulation, parameter estimation and model analysis. Two examples of models of complex regulatory networks in *Escherichia coli* and in *Saccharomyces cerevisiae* are given to demonstrate the usefulness of this approach. It can provide a framework for straightforward development of virtual representations for cellular systems.

INTRODUCTION

Although it is one of the most important challenges in modern biology, a system-level understanding of how cells and organisms function is actually very rudimentary. This is mainly due to the following two reasons: The overwhelming part of experimental investigations can be characterized as qualitative and descriptive, directed towards the understanding of biomolecular details. The concomitant lack of quantitative data will certainly be reduced by further development and wider application of massively parallel experimental methods in functional genomics and proteomics (Uetz et al., 2000; Roberts et al., 2000). Furthermore, due to the complexity of cellular systems even the (nearly) complete measurement of the systems' states *per se* will not enable an integrated understanding of all relevant functional connections and their influence on the observable behaviour (Hartwell et al., 1999).

Recent efforts for a system-level understanding in biology rely on interdisciplinary approaches combining concepts from biology, information sciences and systems engineering. They especially stress the importance of mathematical modeling of complex biological systems in order to come to a virtual representation of cells and organisms. In the end, this representation should allow for computer experiments similar to experiments with real biological systems. Thus systematic testing of biological hypotheses as well as purpose-driven design of cellular functionality are perspectives of these approaches (Hartwell et al., 1999; Stokes, 2000).

The use of mathematical models including the development of computer tools for model formulation and simulation has been demonstrated, for example, by Tomita *et al.* (Tomita et al., 1999) who were able to establish a hypothetical cell comprising 127 genes. Schaff *et al.* (Schaff et al., 1997) follow comparable approaches in the development of a "Virtual Cell" . However, two major challenges for the application of mathematical concepts in the life sciences still have to be resolved: (i) the work on a conceptional framework promoting interdisciplinary research in this direction by finding a "common", non-mathematical language and (ii) a clearly defined modeling concept adapted to cellular systems that allows for easy model development and interpretation (Stokes, 2000).

Focusing on the internal structure of cellular systems, one central, increasingly accepted notion is that these systems are composed of 'functional units' or 'modules'. In this respect, biological systems are more closely related to synthetic, engineered systems than to physical systems (Hartwell et al., 1999; Lauffenburger, 2000). Therefore, a promising way to come to a system-level understanding of cells and organisms is to extend successful theoretical concepts established for the analysis and synthesis of complex technical systems (Gilles, 1998) to biological systems.

On this basis we are currently developing a system- and signal-orientated modeling concept for cellular systems (Kremling et al., 2000; Lengeler, 2000). It relies on the modular structuring of these systems and a systematic representation of biomolecular components in modeling objects. The modeling concept will be outlined in the following section. Afterwards we provide a short sketch of the nature of interdisciplinary research to be carried out in order to establish a "Virtual Biological Laboratory". The usefulness and validity of our approach will be demonstrated by two examples of cellular functional units: the system controlling catabolite repression in *Escherichia coli* and aspects of a complex regulatory network involved in cell cycle regulation in budding yeast.

MODULAR MODELING CONCEPT

The notion of a living cell being composed of subunits of limited autonomy (functional units) plays a prominent role for the modular modeling concept. For the mathematical modeling of cellular systems, this modular structure raises the possibility to independently develop mathematical models for each of the functional units. Hence, submodels as entities in the "model world" correspond to functional units in the "real world". These submodels can later be connected to obtain a description at the system-level. As this approach depends on the identification and representation of functional units, one important question is how to demarcate these units, i.e. how to decompose a complex cellular biochemical network into smaller units. Before we discuss this topic, we will first give a sketch of the overall structure of cellular systems in order to point out broader lines of what to include in the modeling process under our paradigm and then explain how this is done.

General reflections on metabolism and cellular regulation

At a very abstract level, a cell can be divided into two general subnetworks, a regulatory network and a metabolic network (Kremling et al., 2000) as shown in Fig. 9.1. These networks possess very different characteristics: The metabolic network is mainly occupied with substance transformation, e.g. to provide metabolites and cellular structures. In many cases it involves fast biochemical reactions. The regulatory network's main task is information processing, e.g. for the adjustment of enzyme concentrations to the requirements of variable internal and external conditions. This network involves the use of genetic information. Compared to information flow, mass flow only plays a subordinate role in the regulatory network. In this sense, the regulatory network is *superimposed* onto the metabolic network, fulfilling functions analogous to a controller in a technical process.

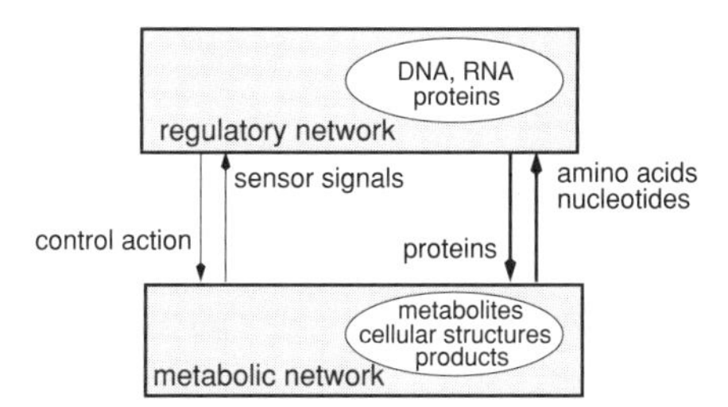

Figure 9.1 Regulatory network and metabolic network: Cellular components constituting the networks and the major connections between them implying signal exchange (left) and substance flow (right).

The interaction between both networks is necessarily bound to substance exchange due to the requirements for precursors and proteins. However, the main connections consist in *directed signal flow*, i.e. sensor signals (e.g. generation of second messengers) and control action (e.g. adjustment of enzyme concentrations). For a system-wide understanding and description of cellular function, these relations between metabolism and regulation imply two major consequences: Firstly, due to their prominent role in bringing about the systems' behaviour, cellular regulation has to be described in a more coherent (and detailed) way than in "traditional" approaches to mathematical modeling. Secondly, more attention has to be given to a signal-oriented view of these systems, which until now have mainly been considered from a mass-flow oriented point of view.

As cellular regulation is established by especially complex gene and protein networks, a closer look at the overall structure of cellular regulation may help to deal with this kind of complexity. In this respect, one important feature of the regulatory network is its *hierarchical structure*. As shown in Fig. 9.2 for transcriptional regulation in budding yeast, the system's possible behaviour on a lower level is constrained by regulation at higher levels. For example, the presence of RNA-Polymerase offers a wide variety of different gene expression patterns, but the actual gene expression is adjusted by combinatorial control involving associated factors and specific transcription factors. Transcription is thus affected by layers above the influence of gene-specific regulators, which enables the cell to establish global to local layers of regulation by controlling the availability of more and more specific components associated with a general transcriptional machinery (Holstege et al., 1998). Similar control structures can be found as a common theme in translation (Sachs and Buratowski, 1997) and in intracellular proteolysis (Kirschner, 1999).

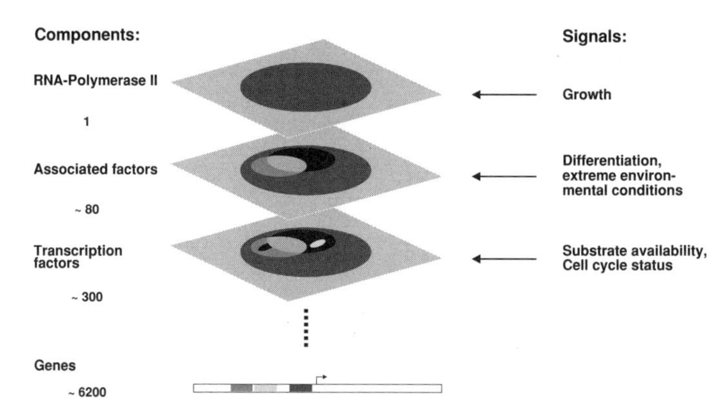

Figure 9.2 Hierarchical structure of the regulatory network: Example of transcriptional regulation in budding yeast. Specificity of regulation increases from global regulation to single gene expression (top to bottom), whereby the components involved (left) become more specific as well as the internal or external signals (right) processed. Shaded areas at each regulatory level indicate the respective behaviour in a system-theoretical sense (Williems, 1989; Willems, 1991) allowed by the combination of all regulatory interactions including higher levels of control.

For the system-wide description of a cell, capturing these constitutive principles in the modeling process has first of all the negative consequence of resulting in detailed and thus seemingly more complex mathematical models. This is compensated for by several advantages as the detailed description enables (i) to consider system-wide coupling of cellular regulation and hence to describe the interplay of global and local control, (ii) to integrate knowledge on well-characterized general components in order to greatly factilitate parameter determination for special subsystems and (iii) to exploit hierarchical network structures for model reduction (Kremling et al., 2000). These aspects will be discussed in more detail in the context of the example systems provided in the following sections.

Identification and representation of functional units

The modular modeling approach presented here depends on the identification and representation of functional units. As we aim at integrating cellular mass- and signal processing functions, each of these functional units has to be composed of a part of the metabolic network and a corresponding part of the regulatory network. Due to their functional dominance (see above), regulatory interactions also have to have a prominent role for the *demarcation of functional units* (or modules).

For this demarcation, we use a preliminary set of three biologically motivated criteria. To be (relatively) self-contained, the modules have

(i) to perform a common physiological task such as represent a linear pathway for amino acid synthesis,

(ii) to be controlled at the genetic level by common regulators i.e. identical transcription factors / the organization in one operon

(iii) and to possess a common information processing (signal transduction) network.

The essential feature of this approach is the combination of classical concepts in the analysis of metabolic systems with a signal-oriented perspective to cellular regulation. Distinct to our approach, several authors adressed the question of demarcation in a more quantitative, flux-oriented way regarding either metabolic pathways (Rohwer et al., 1996; Schuster et al., 2000a; Schilling et al., 2000) or intracellular signal processing networks (Kholodenko et al., 1997; Schuster et al., 2000b). Because systematic investigations on larger modular systems such as the work by van der Gugten *et al.* (Gugten and Westerhoff, 1997) are only at the beginning we use this heuristic way of demarcating functional units. Further work on these theoretical questions will be necessary to come to a more stringent formulation of the criteria cited above.

The application of these criteria enables an entire cellular system to be structured and therefore means a holistic approach to cellular function. Depending on the desired degree of resolution of subsystems, it offers a flexible description of hierarchically nested modules (Fig. 9.3). An enzymatic reaction in glycolysis belongs accordingly to the functional unit "glycolysis" which in turn is part of the larger unit "catabolism".

Our modular modeling approach involves the *systematic representation* of the above identified biological functional units in submodels (modeling objects). At the most fundamental level, a finite and disjunct set of so-called "elementary modeling objects" (Fig. 9.4) has been defined. In Fig. 9.3, for example, each reaction and each storage in the symbolic scheme of glycolysis is described by such an elementary modeling object.

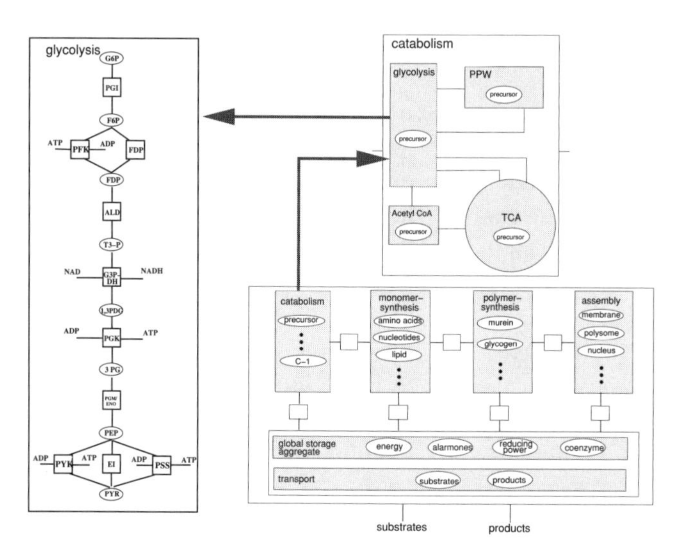

Figure 9.3 Structural decomposition of cellular systems (II): Example for hierarchical nesting of modules

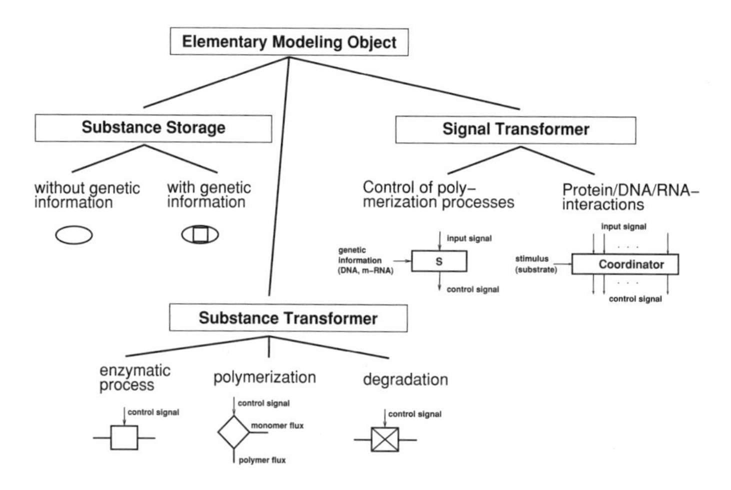

Figure 9.4 Hierarchy of elementary modeling objects for cellular systems. Substance exchange is marked by bold lines whereas arrowheads are used to indicate signal connections.

Elementary modeling objects as well as modeling objects of higher structure are characterized by the following three properties:

(i) They have *structural properties* representing the number and types of inputs and outputs. A simple submodel for an enzymatic reaction, for instance, needs at least two inputs / outputs connected with mass flow of substrate / product and one control signal for enzyme concentration.

(ii) The modeling objects are assigned *behavioural properties*, i.e. mathematical equations describing the dynamic behaviour. Depending on the modeling objectives these equations include algebraic equations, ordinary or partial differential equations (ODEs / PDEs). Often the mathematical equations as the "core" of each modeling object are derived from elementary chemical reaction networks applying chemical kinetic theory (detailed models). To allow for an adjustable degree of model accuracy as well as for efficient simulation, model reduction e.g. via quasi steady-state assumptions is carried out where appropriate. In the case of the single-enzyme example, a Michaelis-Menten kinetic equation could be used.

(iii) Furthermore, each modeling object is assigned a specific *symbolic representation*. Thus even for complicated models (see examples in the following sections) a high degree of biological transparency is guaranteed due to the modular model structure. This is especially important to facilitate interdisciplinary discussions on the underlying biological structures and mechanisms.

The elementary modeling objects are used to represent the basic processes of substance formation (either via simple enzymatic processes or via polymerization processes involving global components like RNA polymerase), degradation and storage. The set of modeling objects is completed by those units representing the corresponding signal transformation processes: Here we define two major subclasses for the control of polymerization processes (transcription / translation) and for the general description of interactions between, for example, proteins. For further details on these elementary modeling objects and their prototypical behavioural characteristics see (Kremling et al., 2000).

Elementary modeling objects can subsequently be interconnected to form higher aggregated structures (Fig. 9.5). A modeling object for transcription comprises polymerization via RNA polymerase and the control of this process. The modeling object for gene expression is obtained by the linkage of transcription and translation submodels.

In summary, the modular modeling approach enables one to progressively obtain a holistic description of more complex functional units. The organization of these modeling objects in an object-oriented class hierarchy also lays the basis for computer-aided model development as described in the next section.

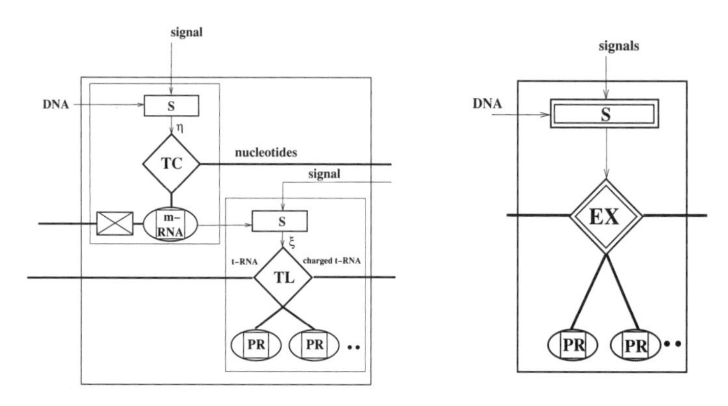

Figure 9.5 Composition of higher structured modeling objects. Elementary modeling objects are linked to a submodel describing gene expression (left). The aggregated model is assigned the representation given on the right hand side.

THE VIRTUAL BIOLOGICAL LABORATORY: AN OUTLINE

One main purpose of the Virtual Biological Laboratory is to enable computer experiments with cellular systems in analogy to experiments carried out with real biological systems in the laboratory. Applications include the quantitative and qualitative analysis of overall behaviour, systematic design of functional units by genetic modifications and the systematic planning of real laboratory experiments. The Virtual Biological Laboratory has to integrate mathematical models with a sound biological background and methods for data storage, computer-aided modeling, simulation and model analysis in a software tool (Fig. 9.6). Accordingly, the development of such a tool requires the close cooperation of biologists, information scientists and system scientists.

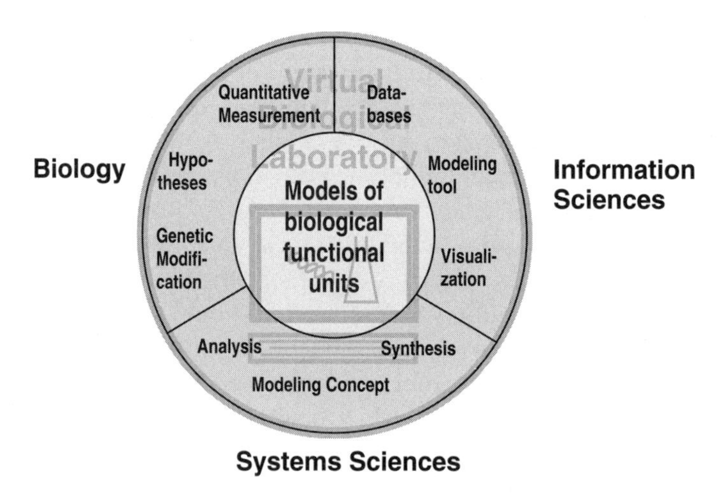

Figure 9.6 Elements to be integrated into a Virtual Biological Laboratory.

The necessity of contributions by each of the three disciplines arises also from the fact that model development has to be understood as an iterative process leading to a maximal convergence of "model world" and "real world". It always requires a careful evaluation of all hypotheses and assumptions by comparison with experimental data. Two of the main tasks of biology are to provide these data and to develop methods for specific perturbation of cellular processes. Information science is needed for database design including a systematic representation of experimental and kinetic data, the development of computer-based modeling tools and finally the implementation of visualization techniques. The system sciences primarily have to provide theoretical methods for demarcation of network structures, system-level analysis and synthesis. General control principles derived from the analysis of cellular networks may also prove beneficial for further development of hierarchical concepts (Raisch et al., 2000) to be applied to control technical systems.

The Virtual Biological Laboratory is currently under development and major parts of it have already been established: The process modeling tool PROMOT, originally designed for application in chemical engineering, allows for the computer-aided development and implementation of mathematical models for living systems (Ginkel et al., 2000). For the numerical analysis of the resulting models the simulation environment DIVA (Mangold et al., 2000) is used. The overall structure of the software is shown in Fig. 9.7.

The modeling methodology of PROMOT distinguishes structural, behavioural and object-oriented modeling. During structural modeling, modules and their interfaces, the so called terminals, are identified according to the biological modeling concept shown in the previous section and aggregated in an aggregation hierarchy of modules. On every level of this hierarchy the modules are linked together using their terminals. These links represent the (possibly bi-directional) exchange of material, momentum, energy or information between the modules.

Modules contain behavioural modeling entities, i.e. variables and equations. They form a differential algebraic equation system (DAE), that is used during the simulation. The description of the behaviour is done in a symbolic way, thus no background knowledge about special numeric algorithms and their implementation is required. PROMOT not only allows for the development of separate models but, as shown in Fig. 9.7, it also enables the implementation and use of flexible, object-oriented knowledge bases containing reusable modeling entities. The structural and behavioural modeling entities are represented as modeling classes that are organized in an inheritance hierarchy. In this hierarchy abstract superclasses are described, which pass on common modeling knowledge to a set of special modeling entities (subclasses). Hence users can not only aggregate modeling entities which are already contained in the knowledge base but they are also able to extend modeling classes by a special

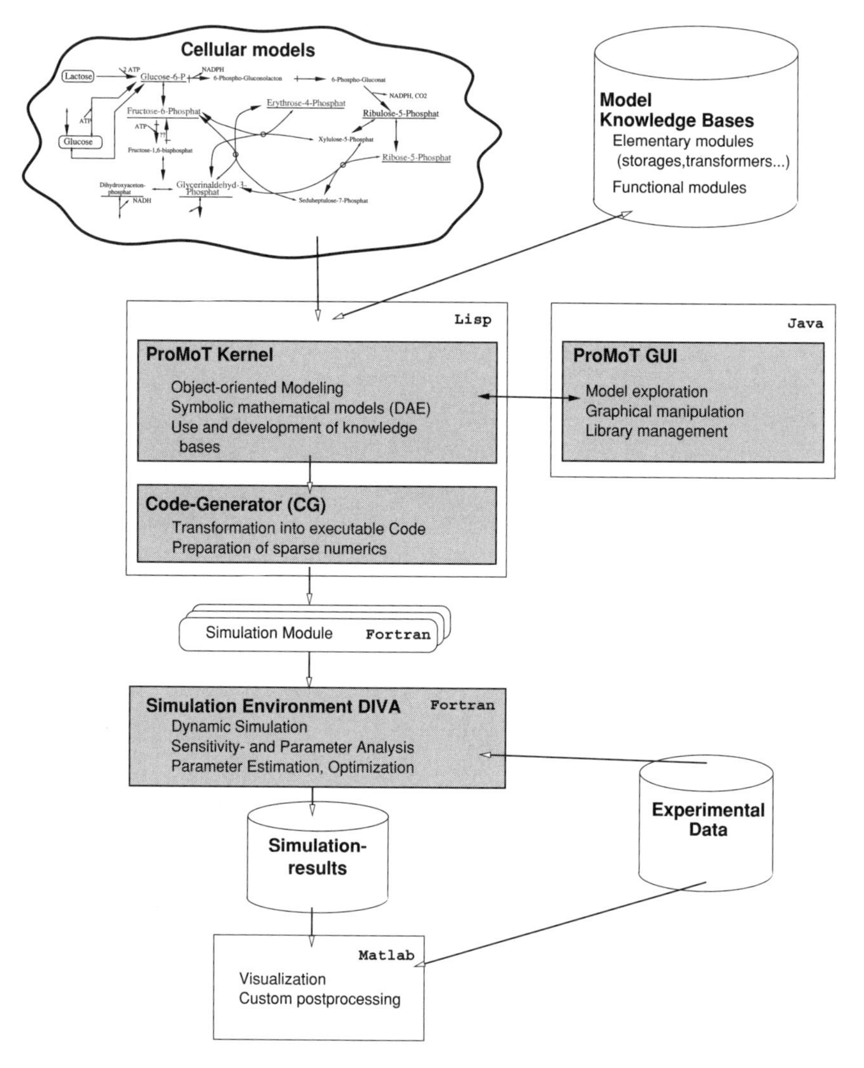

Figure 9.7 Current software architecture of the Virtual Biological Laboratory.

subclass (e.g. to use a special enzyme kinetic) without re-implementing the common parts.

Models in PROMOT can be implemented using the declarative Model Definition Language (MDL) (Tränkle et al., 1997) as well as a graphical user interface (GUI). These two parts can be used alternatingly during model development. The GUI especially supports structural modeling with aggregation and linking of modules and provides a graphical representation whereas MDL is well suited for the description of the behaviour with variables and equations. MDL is also the storage format for the knowledge bases. For the modeling of catabolite repression in *E.coli* (see next section) a knowledge base has been developed which contains the elementary modeling objects described above and some common metabolic subnetworks. An example for a simple metabolic pathway modeled in PROMOT is shown in Fig. 9.8.

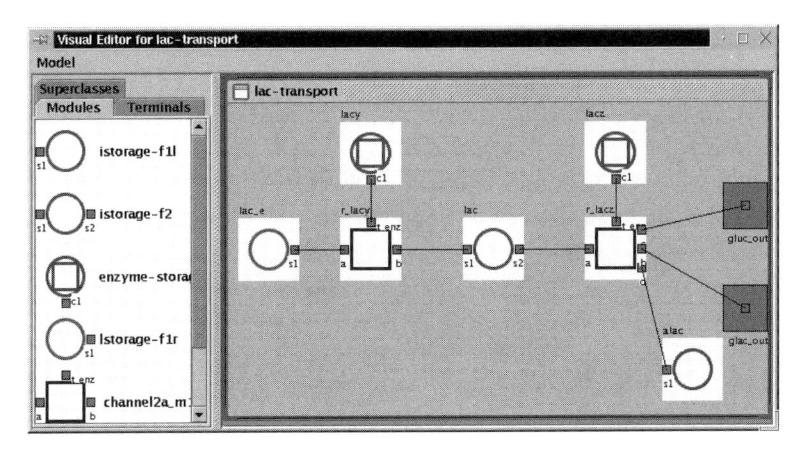

Figure 9.8 Structure of a simple pathway within PROMOT. Lactose is taken up with the help of the enzyme `lacy` and is split into glucose and galactose by `lacz`. An important by-product is Allolactose. The products can be used outside this module by the terminals `gluc_out` and `glac_out`.

Using knowledge bases, model formulation means the selection and linking of pre-defined modeling objects via the GUI and parametrizing the resulting model. The design and implementation techniques in ProMoT are derived from the object-oriented methodology in computer science which was developed for building very complex and flexible software. Using flexible and seperately tested modules it should be possible to build up very large and complex cellular models in a similar way, e. g. a model of a whole cell. With respect to object-oriented and graphically supported model development, the Virtual Biological Laboratory differs significantly from other biological simulation environments such as *E-Cell* (Tomita et al., 1999). Although *E-Cell* also has object-oriented primitive modeling entities like substances and reactions, the users are not able to build their own classes for metabolic networks. Speed and easiness of model development - even without knowing exactly about the underlying mathematical formulations - are thus greatly increased.

Mathematical models generated using PROMOT can be analyzed within the simulation environment DIVA (Mangold et al., 2000, and refs. therein). This simulation tool has been designed especially for dealing with large-scale dynamical (differential-algebraic) systems, which arise in chemical process engineering, but also in the mathematical modeling of complex cellular networks. The model representation for DIVA is compiled from FORTRAN sources (shown as 'Simulation Module' in Fig. 9.7) to machine code. These FORTRAN files are generated by PROMOT using the Code Generator (Köhler et al., 1997). Models in DIVA are handled by sparse-matrix numerics which makes the simulator capable to work on models with up to 5000 differential equations.

Inside DIVA many different numerical computations can be performed based on the same model, including dynamic and steady state simulation, parameter estimation, optimization and the analysis of nonlinear dynamics. There are currently four methods of special interest for cellular models:

(i) Dynamic simulation of the model with different integration algorithms
(ii) Sensitivity analysis for parameters with respect to experimental data
(iii) Parameter identification according to experimental data
(iv) Model-based experimental design.

Most numerical algorithms in DIVA are taken from professional numerical libraries like HARWELL (Harwell Subroutine Library, 1996) and NAG (NAG, 1993) and are therefore very sophisticated. The system offers also additional methods like steady state continuation and bifurcation analysis which are currently only used for process engineering tasks but may also become interesting for cellular systems in the future.

The visualization and postprocessing of the simulation results are done within the standard numeric software MATLAB. This shows one of the drawbacks of the currently available software structure: it consists of rather loosely coupled programs that can exchange data almost only in one direction. To form an efficient workbench (like that depicted in (Kitano, 2000)), the single parts should be integrated more tightly, e. g. to allow the control of the simulation directly from the modeling GUI or to include simulation results as parameters for a simulation module back into PROMOT.

The combination of PROMOT and DIVA is well suited to form the core of the "electronic infrastructure" of a Virtual Biological Laboratory. Examples for the content of the Virtual Biological Laboratory will be given in the following two sections. We will present signal-oriented models for catabolite repression in *E. coli* and for aspects of cell cycle regulation in budding yeast, respectively.

EXAMPLE: CATABOLITE REPRESSION IN *E. COLI*

The expression of carbohydrate uptake systems and metabolizing enzymes is very well controlled in bacteria in order to avoid the useless expression of proteins. For growth, some carbohydrates are preferred to others, resulting in the sequential use of different carbohydrates in mixed cultures. The best examined example of this phenomenon is the diauxic growth of *E. coli* in cultures with glucose and lactose (Neidhardt et al., 1990). Different regulatory proteins contribute to controlling the expression of the corresponding operons and the activity of carbohydrate uptake systems. Being extensively studied over the past few decades, glucose-lactose diauxie of *E. coli* is a perfect model system of complex regulatory networks. Fig. 9.9 gives a survey on the whole model. The symbols for the individual modeling objects can be found in Figs. 9.4 / 9.5. The following functional units are discussed:

Lactose transport and metabolism. The regulatory proteins involved in glucose-lactose diauxie in *E. coli* influence the expression of the lactose metabolizing enzymes. The lactose repressor, LacI, is able to bind to a control sequence in front of the *lac* operon in the absence of lactose, thereby inhibiting transcription from *lacZp*. This repression is relieved in the presence of allolactose, the natural molecular inducer of the *lac* operon.

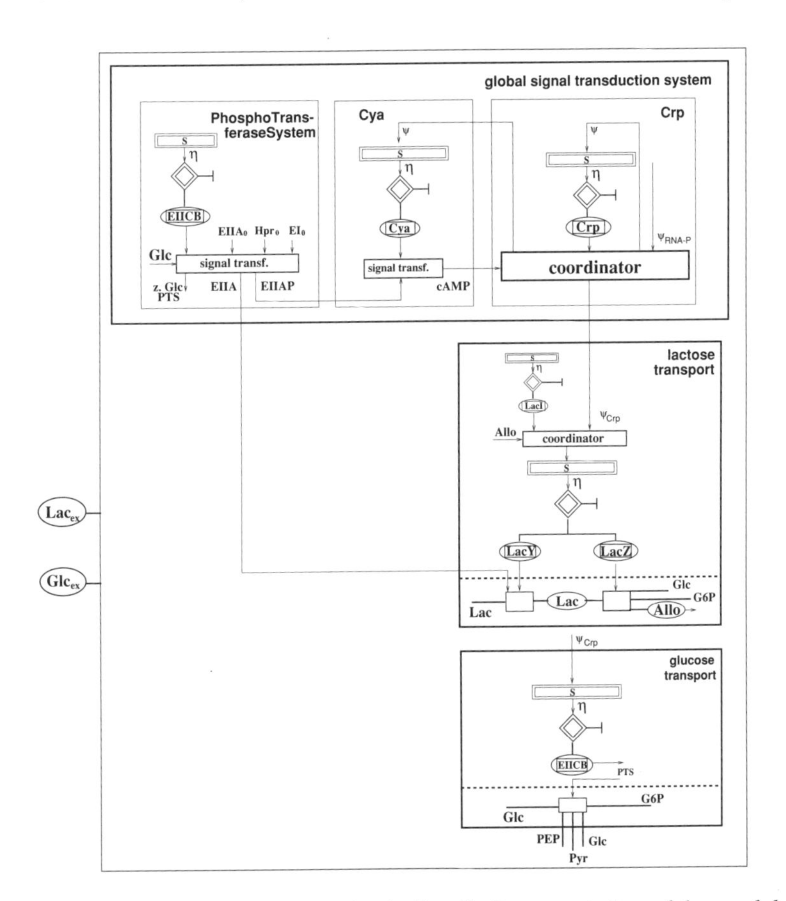

Figure 9.9 Catabolite repression in *E. coli* : Representation of the model structure by using the symbols introduced above. The model comprises lactose and glucose transport and their control by the global signal transduction system *crp* modulon, named after the regulator Crp.

Global signal transduction. Additional control is exerted by the Crp protein. This protein is active in the regulation of a number of operons, most involved in carbohydrate uptake. The Crp protein is able to form a complex with cAMP, that acts as a transcriptional activator for the *lac* operon as well as for the other members of the *crp* modulon (e.g. further glucose uptake systems). The concentration of the alarmone cAMP inside the cell is regulated by complex mechanisms. These mechanisms are basically understood, but despite many well-established details some questions remain. Central in its regulation is the action of the phosphoenolpyruvate-dependent phosphotransferase systems (PTSs), especially the glucose PTS. If the PTSs are not active in the uptake of substrates, the PTS proteins, including Crr which acts as the EIIA in glucose transport, accumulate in their phosphorylated form. Crr$\sim$P is needed for the activation of the enzyme adenylate cyclase (CyaA) that converts ATP into cAMP. Hence, an activation of CyaA is possible only in the absence of PTS substrates or their transport respectively. This leads to an increased level of cAMP inside the cell and in the formation of the cAMP·Crp activator complex. As a result operons like the *lac* operon that depend on the Crp·cAMP complex for transcription can only be expressed if no PTS-substrates are present. Vice versa PTS-substrates in the medium repress transcription of the members of the *crp* modulon. This regulation has therefore been termed catabolite repression.

A mathematical model describing carbon catabolite repression was developed and validated with a set of experiments. The model equations describing signal processing in the global signal transduction system can be found in (Kremling and Gilles, 2000). A further contribution describing the remaining system as well as the experimental setup is in preparation. The model comprises ODEs for the components in the liquid and the biophase. Furthermore algebraic equations are used to describe protein-DNA interactions during gene expression. As can be seen in Fig. 9.9, the model shows a hierarchical structure: The pathways for glucose and lactose are under the control of a superimposed signal transduction pathway.

To validate the model, isogenic mutants, i.e. strains derived from one identical wild type strain with a defined mutation in the signal transduction pathways, were constructed and analyzed. In the experiments the carbohydrate supply in the growth medium as well as in the preculture was varied.

Parameter identification was performed by a two step procedure: (i) Parameter analysis was performed to get those parameters which could be estimated according to the available measurement (see Fig. 9.10). The procedure is based on the calculation of the Fisher information matrix (Posten and Munack, 1990; Ljung, 1999). With measurement data for eight states, 15 of 85 parameters could be estimated together. (ii) Parameter estimation was performed using the SQP (**S**equential **Q**uadratic **P**rogramming) algorithm E04UPF from the NAG library (NAG, 1993). Fig. 9.10 shows the time course of the measured states as well as simulation results (solid lines).

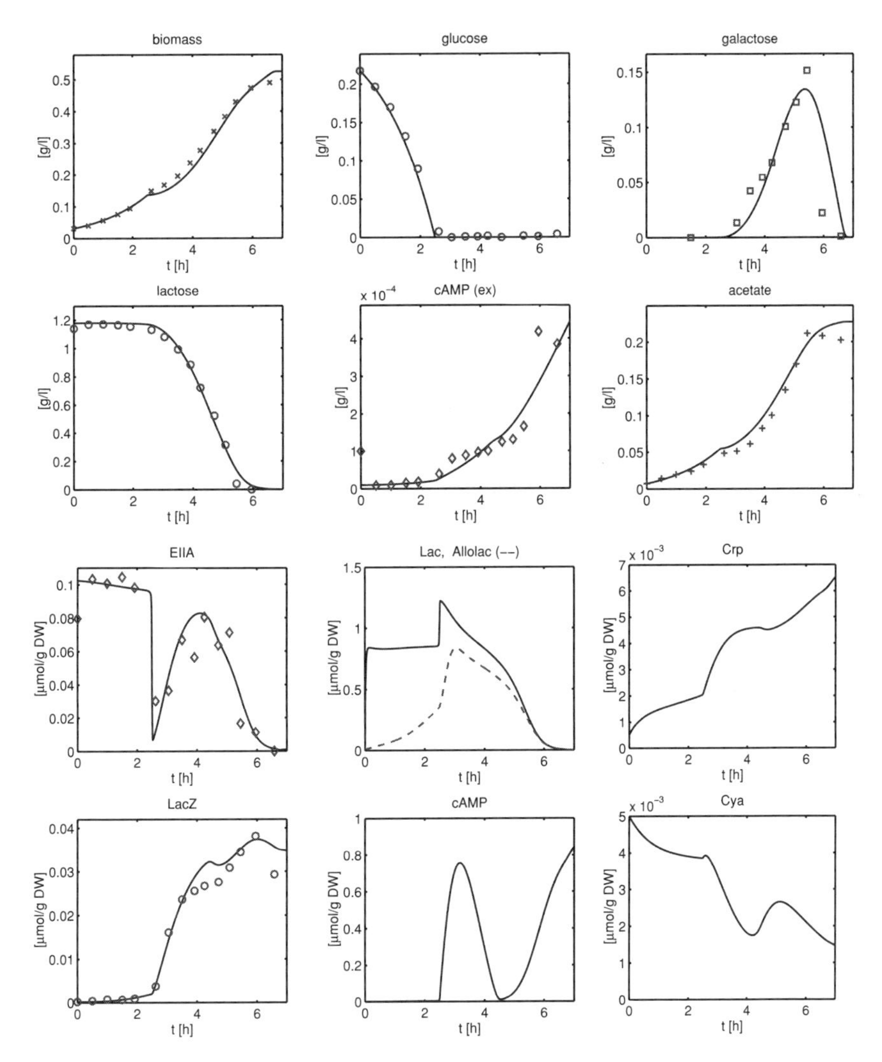

Figure 9.10 Catabolite repression in *E. coli* : Simulation and experimental results for *E. coli* K-12 wildtype. Experiments were carried out with the wildtype strain LJ110 (Zeppenfeld et al., 2000) growing in standard phosphate medium (Tanaka et al., 1967) supplemented with glucose and lactose. Given are the time courses of biomass, glucose, lactose, galactose, cAMP and acetate in the liquid phase. Measurement of intracellular EIIA and LacZ as well as simulation of further intracellular states are also shown (open symbols and '+' indicate measurement; solid lines indicate simulation results). Quantification of carbohydrates and acetate in the growth medium was carried out by using the respective testkits of Roche Diagnostics GmbH (Germany) according to the instructions of the manufacturer. Determination of β-galactosidase activity was done as described by (Pardee and Prestige, 1961). Determination of phosphorylation state of Crr was carried out essentially as described by (Hogema et al., 1998). The amount of extracellular cAMP was determined with the Cyclic AMP Enzyme Immunoassay Kit of Cayman Chemical (Ann Arbor, USA).

To summarize the results of this model, it can be stated that (i) the model quantitatively describes experimental results obtained with a number of mutant strains, (ii) the model allows the prediction of the time course of not yet measurable variables like cAMP, (iii) the model can be used as a basis for further analysis. It is now used to test hypothesises about regulatory phenomena influencing the growth of some mutant strains.

EXAMPLE: CELL CYCLE REGULATION IN BUDDING YEAST

In all eukaryotic cells the cell division cycle is characterized by a fixed sequence of cell cycle phases (Fig. 9.11), during which the main cellular tasks are switched from simple mass growth (G1 phase) to DNA replication (S phase) and finally to chromosome separation and cytokinesis (G2 / M phase). In response to multiple internal and external signals, the sequence is mainly controlled by cyclin dependent kinases (CDKs). They are activated by phase-specific cyclins forming distinct kinase complexes with different functionality.

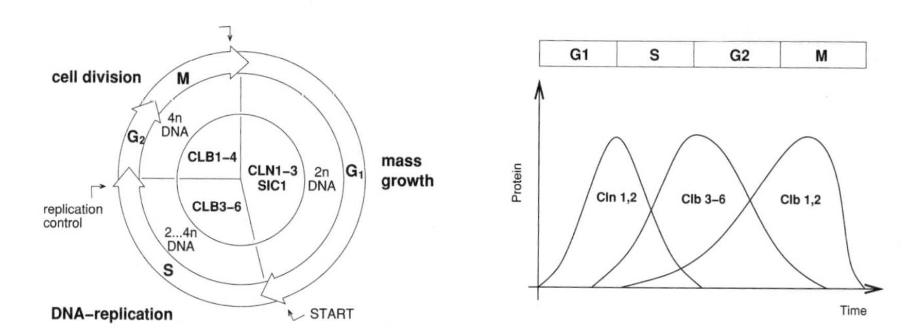

Figure 9.11 Cell cycle regulation in *Saccharomyces cerevisiae* : Main cellular tasks in specific cell cycle phases, checkpoints, DNA-content, cell cycle regulators (left) and a schematic view of the corresponding cyclin concentrations (right).

Even in the relatively simple yeast *Saccharomyces cerevisiae*, one catalytic subunit (Cdc28) and nine cyclins (Cln1-3 / Clb1-6) are involved in cell cycle regulation (Mendenhall and Hodge, 1998). The phase-specific cyclin fluctuation in this organism relies upon such diverse processes as regulated transcription of cyclin genes, constitutive or controlled protein degradation and specific inactivation of Clb-CDKs via the CDK inhibitor Sic1. All regulators are embedded in a highly interconnected network including positive and negative feedback loops (Mendenhall and Hodge, 1998). Additionally, cell cycle regulation in budding yeast not only serves as an example for a complex regulatory network; it also involves many of the known regulatory mechanisms at the DNA, mRNA and protein levels which generally have to be accounted for during model development.

In the cell cycle, the G1/S-transition plays a crucial role, because at this boundary – via the associated checkpoint called "START" – the cells ultimately have to decide whether to undergo a new round of replication and division or not. The accumulation of sufficient cellular material, i.e. the attainment of a critical cell size, constitutes the major prerequisite for this transition (Mendenhall and Hodge, 1998). At the molecular level, the transition is governed by an approximately constant level of Cln3, which surprisingly results in the sudden activation of a transcription factor composed of Swi4 / Swi6. In this way the production of G1 cyclins Cln1/2 induces the transition to the S phase. Whereas these regulatory mechanisms are well established, finding a consistent explanation for the sudden appearance of G1 cyclins as a function of cellular growth is complicated (Mendenhall and Hodge, 1998).

To quantitatively analyze the system's dynamics, a submodel was formulated according to the modeling concept outlined above. Its structure, which is based solely on the known regulatory mechanisms, is shown in Fig. 9.12. Special attention was given to incorporate the interaction between regulatory processes at the DNA as well as at the protein level.

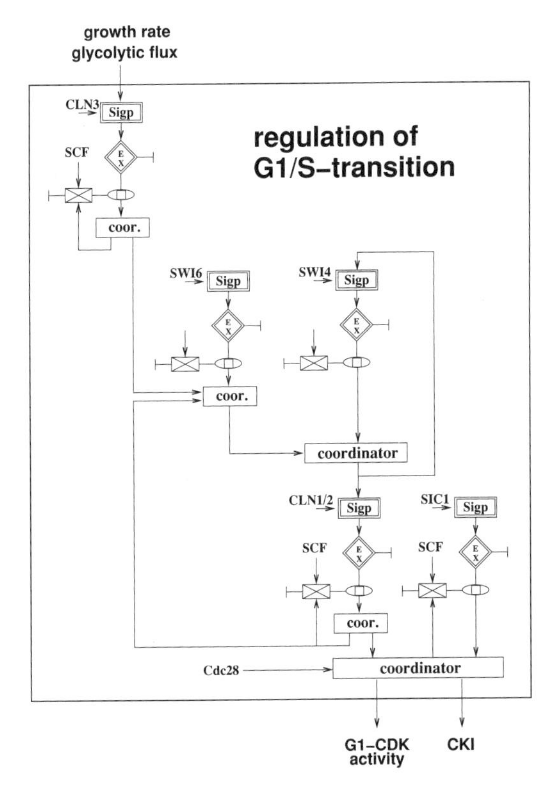

Figure 9.12 Model structure capturing the regulatory network for the control of the G1 / S-transition during the cell cycle in *Saccharomyces cerevisiae*. Symbols represent modeling objects as defined in Figs. 9.4 / 9.5.

Unfortunately, in the area of cell cycle regulation kinetic parameters and absolute concentrations are available only for specific cases like the Swi4-promoter interaction (Taylor et al., 2000). According to the modeling concept outlined above, we therefore first used quantitative data from literature concerning global components like RNA polymerase II (Thomas et al., 1997) to constrain the possible model behaviour to the range found *in vivo*. Further parameter values were obtained by taking into account structural characteristics (e.g. gene lengths) and data from microarray experiments for (relative) concentration profiles (Cho et al., 1998) as well as for mRNA stability (Holstege et al., 1998). Based on experimental data covering single mechanistic aspects like (Yaglom et al., 1995) and connections between growth rate and G1 length (Aon and Cortasse, 1999), the order of magnitude of the remaining parameters was estimated. Parameter analysis finally revealed the submodel as being *robust*, i.e. being relatively insensitive to the precise values of model parameters with respect to its qualitative behaviour (see below). As with other complex regulatory networks (Alon et al., 1999; von Dassow et al., 2000), this seems to be a direct consequence of the network's architecture. It also gives a first hint at the correctness of the model structure.

Several conclusions regarding the character of the G1 / S transition can be drawn from the simulation results (Fig. 9.13): Although held at an approximately constant concentration, Cln3 is able to drive the transition as a function of cellular growth. Mechanistically, the control of *CLN1/2* transcription via Swi4/6 plays a prominent role in this process. Due to several positive and negative feedback loops, the system behaves as a switch function as soon as a Cln3 threshold is reached. The known regulatory mechanisms therefore sufficiently explain the behaviour observed *in vivo*.

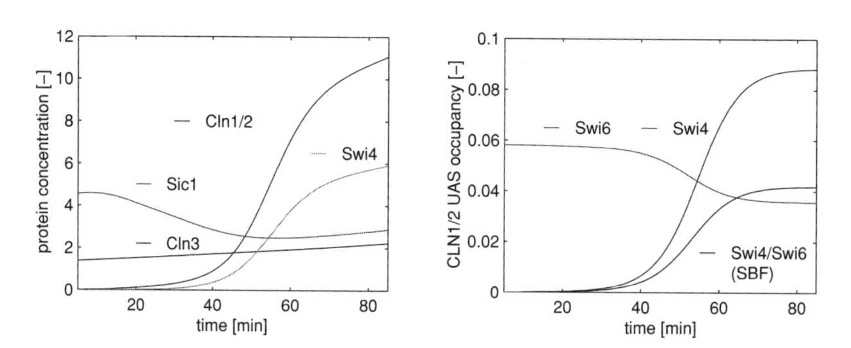

Figure 9.13 Simulation results for regulation of the G1 / S transition: Protein concentrations of G1 cyclins, CDK inhibitor and transcription factor Swi4 (left) and transcriptional regulation of the *CLN1/2* genes (right).

Differing from a published mathematical model of cell cycle regulation in budding yeast (Chen et al., 2000), our (partial) model is based on the deduction of dynamic properties from a more detailed description of regulatory mechanisms: Without implementing, for example, an ultrasensitive switch function for the activation of the transcription factor SBF, the behaviour results from an interplay of regulated gene expression, phosphorylation / dephosphorylation reactions and cooperative binding to multiple sites on the DNA.

To summarize, the major biological findings obtained by this model are: (i) biological knowledge on the regulatory network under consideration is sufficient to explain the observed behaviour, (ii) the seemingly complex behaviour results from the interplay of regulatory circuits, which have to be viewed in a quantitative way to get a clue on the entire network's function and (iii) mathematical modeling gives hints that this network constitutes a relatively robust regulatory module.

CONCLUSIONS

Finding concepts to deal with the complexity of living systems represents the major challenge on the way to a system-level understanding of cells and organisms. In this contribution, we present a framework which is derived from concepts in engineering science and systems theory. It essentially relies upon the modular mathematical modeling of the overall behaviour of cellular functional units. The decomposition of cells into such units is oriented at the modular biomolecular structure of cellular systems. This demarcation also represents the most crucial aspect of the modeling concept as mainly heuristic criteria are applied at the moment. In the future, theoretical work in this area will be intensified. Finally, the modeling concept should guarantee a high degree of biological transparency and promote the interdisciplinary cooperation between biologists and system engineers.

A long-term perspective of our work is the establishment of a Virtual Biological Laboratory combining mathematical models of cellular systems with tools for their efficient development, simulation and analysis. The purpose of this laboratory is to enable computer experiments with cellular systems similar to the analysis and design of cellular systems in the "real" world. One milestone in reaching this aim is the development of process modeling and simulation tools forming the "core" of the Virtual Biological Laboratory. As we have shown, computational tools like PROMOT and DIVA are already available and allow for a straight-forward realization of the modular modeling concept outlined in this contribution. For two small example systems, we showed a systematic formulation of mathematical models based on (structural) biological knowledge. This can lead to an adequate description of experimentally observable cellular behaviour and to new insight into how cellular biochemical networks operate.

For the system-level description of more complex systems, even of a simple bacterium like *E. coli*, an intensified cooperation of biology, information sciences and systems sciences is essential. However, with regard to the modular structure of cellular systems, a virtual representation of real cellular systems could be achieved through cooperation and division of labour.

ACKNOWLEDGMENTS

Financial support by a "Peter und Traudl Engelhorn Stiftung" grant to J.S. is gratefully acknowledged.

References

Advanced Computing Department, AEA Industrial Technology, Harwell Laboratory, Oxfordshire, England. (1996). *Harwell Subroutine Library..*

Alon, U., Surette, M.G., Barkai, N., and Leibler, S. (1999). Robustness in bacterial chemotaxis. *Nature* 397:168–71.

Aon, M.A. and Cortasse, S. (1999). Quantitation of the effects of disruption of catabolite (de)repression genes on the cell cycle behavior of *Saccharomyces cerevisiae*. *Curr. Microbiol.* 38(1):57–60.

Chen, K.C., Csikasz-Nagy, A., Gyorffy, B., Val, J., Novak, B., and Tyson, J.J. (2000). Kinetic analysis of a molecular model of the budding yeast cell cycle. *Mol. Biol. Cell* 11:369–391.

Cho, R.J., Campbell, M.J., Winzeler, E.A., Steinmetz, L., Conway, A., Wodicka, L., Wolfsberg, T.G., Gabrielian, A.E., Landsmann, D., and Davis, R.W. (1998). A genome-wide transcriptional analysis of the mitotic cell cycle. *Mol. Cell* 2:65–73.

Gilles, E.D. (1998). Network theory for chemical processes. *Chem. Eng. Technol.* 21:121–132.

Ginkel, M., Kremling, A., Tränkle, F., Gilles, E.D., and Zeitz, M. (2000). Application of the process modeling tool PROMOT to the modeling of metabolic networks. In I. Troch and F. Breitenecker, editors, *Proc. 3rd MATHMOD, Vienna*, volume 2 of *ARGESIM Report 15*, pp.525–528.

Hartwell, L.H., Hopfield, J.J., Leibler, S., and Murray, A.W. (1999). From molecular to modular cell biology. *Nature* 402 (Supp.):C47–C52.

Hogema, B.M., Arents, J.C., Bader, R. Eijkemans, K., Yoshida, H., Takahashi, H., Aiba, H., and Postma, P.W. (1998). Inducer exclusion in *Escherichia coli* by non-PTS substrates: the role of PEP to pyruvate ratio in determining the phosphorylation state of enzyme IIA^{Glc}. *Mol. Microbiol.* 30:487–498.

Holstege, F.C.P., Jennings, E.G., Wyrick, J.J., Lee, T.I., Hengartner, C.J., Green, M.R., Golub, T.R., Lander, E.S., and Young, R.A. (1998). Dissecting the regulatory circuitry of a eukaryotic genome. *Cell* 95:717–728.

Kholodenko, B.N., Hoek, J.B., Westerhoff, H.V., and Brown, G.C. (1997). Quantification of information transfer via cellular signal transduction pathways. *FEBS Letters* 414:430–434.

Kirschner, M. (1999). Intracellular proteolysis. *Trends Cell Biol.* 24(12):M42–M45.

Kitano, H. (2000). Perspectives on Systems Biology. In *Proceedings of the First International Conference on Systems Biology*, Japan.

Köhler, R., Räumschüssel, S. and Zeitz, M. (1997). Code generator for implementing differential algebraic models used in the process simulation tool DIVA. In A. Sydow, editor, *15th IMACS World Congress*, volume 3, pp.621–626, Berlin, Germany.

Kremling, A. and Gilles, E.D. (2000). The organization of metabolic reaction networks: II. Signal processing in hierarchical structured functional units. *Metabolic Eng.* in press.

Kremling, A., Jahreis, K. Lengeler, J.W., and Gilles, E.D. (2000). The organization of metabolic reaction networks: A signal-oriented approach to cellular models. *Metabolic Eng.* 2(3):190–200.

Lauffenburger, D.A. (2000). Cell signaling pathways as control modules: Complexity for simplicity? *Proc. Natl. Acad. Sci. USA* 97(10):5031–5033.

Lengeler, J.W. (2000). Metabolic networks: a signal-oriented approach to cellular models. *Biol. Chem.* 381:911–920.

Ljung, L. *System Identification – Theory for the user*. Prentice Hall PTR, Upper Saddle River, New Jersey, 2nd edition.

Mangold, M., Kienle, A., Mohl, K.D., and Gilles, E.D. (2000). Nonlinear computation using DIVA - methods and applications. *Chem. Eng. Sci.* 55:441–454.

Mendenhall, M.M. and Hodge, A.E. (1998). Regulation of Cdc28 cyclin-dependent protein kinase activity during the cell cycle of the yeast *Saccharomyces cerevisiae*. *Microbiol. Mol. Biol. Rev.* 62(4):1191–1243.

NAG Ltd., Oxford, England, UK. *NAG Fortran Library Manual.*.

Neidhardt, F.C., Ingraham, J.L., and Schaechter, M. (1990). *Physiology of the bacterial cell: A molecular approach*. Sinauer Associates, Sunderland, Massachusetts.

Pardee, A.B. and Prestige, L.S. (1961). The initial kinetics of enzyme induction. *Biochem. Biophys. Acta* 49:77–88.

Posten, C. and Munack, A. (1990). On-line application of parameter estimation accuracy to biotechnical processes. In *Proceedings of the American Control Conference*, volume 3, pp.2181–2186.

Raisch, J., Itigin, A., and Moor, T. (2000). Hierarchical control of hybrid systems. In S. Engell, S. Kowalewski, and J.Zaytoon, editors, *Proceedings of ADPM 2000*, pp.67–72, Aachen, Germany.

Roberts, C.J., Nelson, B., Marton, M.J., Stoughton, R., Meyer, M.R., Bennett, H.A., He, Y.D., Dai, H., Walker, W.L., Hughes, T.R., Tyers, M., Boone, C., and Friend, S.H. (2000). Signaling and circuitry of multiple MAPK pathways revealed by a matrix of global gene expression profiles *Science* 287:873–880.

Rohwer, J.M., Schuster, S., and Westerhoff, H.V. (1996). How to recognize monofunctional units in a metabolic system. *J. theor. Biol.* 179:213–228.

Sachs, A.B. and Buratowski, S. (1997). Common themes in translational and transcriptional regulation. *Trends Biochem. Sci.* 22:189–192.

Schaff, J., Fink, C.C., Slepchenko, B., Carson, J.H., and Loew, L.M. (1997). A general computational framework for modeling cellular structure and function. *Biophys. J.* 73:1135–1146.

Schilling, C.H., Letscher, D., and Palsson, B.O. (2000). Theory for the systemic definition of metabolic pathways and their use in interpreting metabolic function from a pathway-oriented perspective. *J. theor. Biol.* 203:229–248.

Schuster, S., Fell, D.A., and Dandekar, T. (2000). A general definition of metabolic pathways useful for systematic organization and analysis of complex metabolic networks. *Nature Biotechnol.*, 18:326–332.

Schuster, S., Kholodenko, B.N., and Westerhoff, H.V. (2000). Cellular information transfer regarded from a stoichiometry and control analysis perspective. *Biosystems* 55(1-3):73–81.

Stokes, C.L. (2000). Biological systems modeling: Powerful discipline for biomedical e-R&D. *AIChE J.* 46:430–433.

Tanaka, S., Lerner, S.A., and Lin, E.C.C. (1967). Replacement of a phosphoenolpyruvate-dependent phosphotransferase by a nicotinamide adenin dinucleotide linked dehydrogenase for the utilization of manitol. *J. Bacteriol.* 93:642–648.

Taylor, I.A., McIntosh, P.B., Pala, P., Treiber, M.K., Howell, S., Lane, A.N., and Smerdon, S.J. (2000). Characterization of the DNA-binding domains from the yeast cell-cycle transcription factors Mbp1 and Swi4. *Biochemistry* 39:3943–3954.

Thomas, M., Chédin, S., Charles, C., Riva, M., Famulok, M., and Sentenac, A. (1997). Selective targeting and inhibition of yeast RNA polymerase II by RNA aptamers. *J. Biol. Chem.* 272(44):27980–27986.

Tomita, M., Hashimoto, K., Takahashi, K., Shimizu, T.S., Matsuzaki, Y., Miyoshi, F., Saito, K., Tanida, S., Yugi, K., Venter, J.C., and Hutchinson, C.A. (1999). E-CELL: software environment for whole-cell simulation. *Bioinformatics* 15(1):72–84.

Tränkle, F., Gerstlauer, A., Zeitz, M., and Gilles, E.D. (1997). The Object-Oriented Model Definition Language MDL of the Knowledge-Based Process Modeling Tool PROMOT. In A. Sydow, editor, *15th IMACS World Congress*, volume 4, pp.91–96, Berlin.

Uetz, P., Cagney, G., Mansfield, T.A., Judson, R.S., Knight, J.R., Lockshon, D., Narayan, V., Srinivasan, M., Pochart, P., Qureshi-Emili, A., Li, Y., Godwin, B., Conover, D., Kalbfleisch, T., Vijayadamodar, G., Yang, M., Johnston, M., Fields, S., and Rothberg, J.M. (2000).A comprehensive analysis of protein-protein interactions in *Saccharomyces cerevisiae*. *Nature* 403:623–627.

van der Gugten, A.A. and Westerhoff, H.V. (1997). Internal regulation of a modular system: the different faces of internal control. *BioSystems* 44:79–106.

von Dassow, G., Meir, E., Munro, E.M., and Odell, G.M. (2000). The segment polarity network is a robust developmental module. *Nature* 406:188–192.

Willems, J.C. (1989). Models for dynamics. *Dynamics Reported* 2:171–269.

Willems, J.C. (1991). Paradigms and puzzles in the theory of dynamical systems. *IEEE Trans. Automat. Control* 36(3):259–294.

Yaglom, J., Linskens, M.H.K., Sadis, S., Rubin, D.M., Futcher, B., and Finley, D. (1995). $p34^{Cdc28}$-mediated control of Cln3 cyclin degradation. *Mol. Cell. Biol.* 15(2):731–741.

Zeppenfeld, T., Larisch, C., Lengler, J.W., and Jahreis, K. (2000). Glucose transporter mutants of *Escherichia coli* K-12 with changes in substrate recognition of $IICB^{Glc}$ and induction behavior of the *ptsG* gene. *J. Bacteriol.* 182:4443–4452.

10 Computational Cell Biology — The Stochastic Approach

Thomas Simon Shimizu and Dennis Bray

Although the need for computational modelling and simulation in cell biology is now widely appreciated, existing methods are still inadequate in many respects. The conventional approach of representing biochemical reactions by continuous, deterministic rate equations, for example, cannot easily be applied to intracellular processes based on multiprotein complexes, or those that depend on the individual behaviour of small numbers of molecules. Stochastic modelling has emerged in recent years as an alternative, and physically more realistic, approach to phenomena such as intracellular signalling and gene expression. We have been using and developing a stochastic program, StochSim as a tool to investigate the molecular details of the bacterial chemotaxis signalling pathway. In this chapter, we briefly review the StochSim algorithm and provide a comparison with another popular stochastic approach developed by Gillespie. We then describe our current efforts in extending StochSim to include a spatial representation and conclude by considering directions for future development.

INTRODUCTION

In broad terms, the current interest in "computational cell biology" reflects the contemporary fascination with electronic networks of all kinds. There is a widespread feeling that the speed of computers and the sophistication of programmers can at last match the bewildering molecular complexity of living cells. This viewpoint is encouraged and fed by recent genomic studies, which have built up such an impetus that they are now pushing into areas outside the genome. Evidently, sequence information by itself cannot explain the functioning of a living cell or organism. It is also true that we have built up, over the past century, an enormous body of information about proteins and other molecules inside cells. Why should we not — the argument goes — store, collate and analyze these data by comprehensive, computer-intensive techniques similar to those currently employed to analyze genomes?

Unfortunately, no consensus presently exists as to how best to perform this analysis, or to what we can expect as a result. One clear-cut function of computers in cell biology is to store large amounts of information in a logical and accessible form. This role has seen its most public triumph in the generation of genomic databases, but also underpins the giant strides made in the determination of protein structure. Many databases containing integrated information on specific organisms have been developed and some of these allow the user to access specific molecular details on individual cells, such as WormBase (Stein et al., 2001) and EcoCyc (Karp et al., 2000). There are also large programs containing information about specific cell types, such as the software developed by Denis Noble to analyze the behavior of heart muscle cells (Noble, 2001). Several ambitious projects have been initiated that aim to simulate entire cells or parts of cells at a molecular level, such as E-CELL (Tomita et al., 1999) and the Virtual Cell (Schaff et al., 1997).

Computers were of course used by biologists before the genomic era. The area of metabolic modeling, for example, with its roots in enzyme kinetics, was one of the earliest to be adapted to computer simulation. Many software packages have been written that allow the kinetic performance of enzyme pathways to be represented and evaluated quantitatively, such as GEPASI (Mendes, 1993), MIST (Ehlde and Zacchi, 1995) and SCAMP (Sauro, 1993). This is also an area of commercial interest and biotechnology companies engaged in the production of food or drugs by fermentation or allied processes routinely evaluate their production by flux-analysis programs, often aided by metabolic control analysis.

Neurobiology is another computationally rich area. Whether because of their background in the physical science, or because of the computer-like nature of the brain, neurophysiologists have always been much more open to the use of computers than cell biologists. Several large computer packages have been developed and (what is far more significant) widely used as adjuncts to research neurophysiology. Packages such as GENESIS (Wilson et al., 1989) and NEURON (Hines, 1993) provide integrated suites of routines for the recording and analysis of electrical data, the simulated performance of individual axons, and the investigation of networks of nerve cells and cortical activity.

In contrast to the above areas, topics that come under the rubric of core cell biology — those not directly concerned with DNA sequences, ions, or low molecular weight metabolites — are more difficult to handle computationally. Cell signalling, cell motility, organelle transport, gene transcription, morphogenesis and cellular differentiation cannot easily be accommodated into existing computational frameworks. Attempts to use computers in these areas are still at a stage of exploratory software development, usually in the hands of individual research groups. Conventional approaches using the numerical integration of continuous, deterministic rate equations sometimes provide a convenient route, especially when systems are very large or when molecular details are of little importance. But as the resolution of experimental techniques increases, so the limitations of conventional models become more evident. Difficulties include the combinatorial explosion of large numbers of different species, the importance of spatial location within the cell, and the instability associated with reactions between small numbers of molecular species.

STOCHASTIC SIMULATION OF CELLULAR PROCESSES

In recent years, a number of research groups have attempted to use a radically different approach to the processes occurring within cells (McAdams and Arkin, 1997; Stiles et al., 1998; Morton-Firth and Bray, 1998). The idea is to represent individual molecules rather than the concentrations of molecular species, and to apply Monte Carlo methods to predict their interactions. Motivation for this new approach comes from the realization that many crucial events in living cells depend on the interaction of small numbers of molecules and hence are sensitive to the underlying stochasticity of the reaction processes. Under these conditions, the usual approach taken to biochemical reactions of analyzing their characteristic continuous, deterministic rate equations breaks down and fails to predict the behavior of the system accurately. Signalling pathways, for example, commonly operate close to points of instability and frequently employ feedback and oscillatory reaction networks that are sensitive to the operation of small numbers of molecules (Hallett, 1989; Goldbeter, 1996). Only 200 K^+ and Na^+ channels responsive to changes in intracellular Ca^{2+} are responsible for a key step in many neutrophil signalling pathways (Hallett, 1989). Gene transcription is controlled by small assemblies of proteins operating in an all-or-none fashion, so that whether a specific protein is expressed or not is, to some extent, a matter of chance (Ko, 1991; Kingston and Green, 1994; Tjian and Maniatis, 1994; McAdams and Arkin, 1999). The performance of sensory detectors such as retinal rod outer segments (Lamb, 1994; Van Steveninck and Laughlin, 1996) and even the firing of individual nerve cells (Smetters and Zador, 1996; White et al., 1999) are intrinsically stochastic.

In the stochastic modeling approach, rate equations are replaced by individual reaction probabilities and the output has a physically-realistic stochastic nature. Techniques are available by which large numbers of related species can be coded in an economical fashion and key concepts, such as signalling complexes and the thermally-driven flipping of protein conformations, can be embodied into the program. Stochastic modeling may help us to integrate biochemical and thermodynamic data in a coherent and manageable way.

MODELING BACTERIAL CHEMOTAXIS

We have used both deterministic and individual-based stochastic programs to investigate the pathway of intracellular signals used by coliform bacteria in the detection of chemotactic stimuli (Bray et al., 1993; Bray and Bourret, 1995; Morton-Firth, 1998)[1]. The models are based on physiological data collected from single tethered bacteria of over 60 mutant genotypes. Quantitative discrepancies between computer simulations and experimental data throw a spotlight on areas of uncertainty in the signal transduction pathway, highlighting the importance of spatial organization to the logical operation of the pathway. In particular they emphasize the function of a specific, well-characterized, cluster of proteins associated with the chemotaxis receptors which acts like a self-contained computational cassette.

The individual-based stochastic simulation program STOCHSIM was written by Carl Firth as part of his PhD work at the University of Cambridge (Morton-Firth, 1998). It was developed as part of a study of bacterial chemotaxis to be a more realistic way of representing the stochastic features of this signalling pathway and also as a means to handle the large numbers of individual reactions encountered (Morton-Firth, 1998; Morton-Firth et al., 1999). The program provides a general-purpose biochemical simulator in which each molecule or molecular complex in the system is represented as an individual software object. Reactions between molecules occur stochastically, according to probabilities derived from known rate constants. An important feature of the program is its ability to represent multiple post-translational modifications and conformational states of protein molecules.

DESCRIPTION OF THE STOCHSIM ALGORITHM

In STOCHSIM, each molecule (not each population of molecular species) is represented as an individual software object, and a number of dummy molecules, or "pseudo-molecules", are also included in the reaction system. Time is quantized into a series of discrete, independent time-slices, the size of which is determined by the most rapid reaction in the system. In each time-slice, STOCHSIM first chooses one molecule at random from the population of "real" molecules, and then makes another selection from the entire population including the pseudo-molecules. If two molecules are selected, they are tested for all possible bimolecular reactions for the particular reactant combination. If one molecule and one pseudo-molecule are chosen, the molecule is tested for all possible unimolecular reactions it can undergo.

Reaction probabilities are pre-computed at initialization time, and stored in a look-up table so that they need not be calculated during the execution of each time-slice. These probabilities scale linearly with the size of the time-slice (Eqs. 10.1 and 10.2 below), so that if the time-slices are suf-

---

1 A resumé of this work together with a list of published references can be found at `http://www.zoo.cam.ac.uk/comp-cell`.

ficiently small, a single random number can be used to test for all possible reactions that a particular combination of reactants can undergo. Once the reactant molecules are chosen, the set of possible reactions are retrieved from the look-up table with their probabilities. STOCHSIM then iterates through these reactions in turn and computes a "cumulative probability" for each of the possible outcomes. The set of cumulative probabilities can then be compared with a single random number to choose which reaction, if any, occurs. If a reaction does occur, the system is updated accordingly and the next time-slice begins with another pair of molecules being selected.

The probabilities stored in the look-up table are calculated from the following five parameters: (i) the deterministic rate constant (k_1 and k_2 for uni- and bi-molecular reactions, respectively), (ii) the size of the time increment (Δt), (iii) the number of molecules in the system (n), (iv) the number of pseudo-molecules in the system (n_0), and (v) the volume of the system (V). Using these parameters, the probabilities for uni- and bi-molecular reactions (p_1 and p_2, respectively) are obtained by:

$$p_1 = \frac{k_1 n(n + n_0)\Delta t}{n_0} \tag{10.1}$$

and

$$p_2 = \frac{k_2 n(n + n_0)\Delta t}{2N_A V}. \tag{10.2}$$

The previously published derivation of these expressions (Morton-Firth and Bray, 1998; Morton-Firth, 1998) is summarized in Appendix A for reference.

Whenever a molecular species in the system can exist in more than one state, then the program encodes it as a "multistate molecule" with a series of binary flags. Each flag represents a state or property of the molecule, such as a conformational state, the binding of ligand, or covalent modification (e.g. phosphorylation, methylation, etc.). The flags specify the instantaneous state of the molecule and may modify the reactions it can participate in. For instance, a multistate molecule may participate in a reaction at an increased rate as a result of phosphorylation, or fail to react because it is in an inactive conformation. The flags themselves can be modified in each time step as a result of a reaction, or they can be instantaneously equilibrated according to a fixed probability. The latter tactic is used with processes such as ligand binding or conformational change that occur several orders of magnitude faster than other chemical reactions in the system.

Let us say that in a particular time step, STOCHSIM has selected one or more multistate molecules. It then proceeds in the following manner. First any rapidly-equilibrated "fast flags" on the molecule are assigned to be on or off according to a weighted probability. A protein conformation flag, for example, can be set to be active or inactive, according to which other flags of the molecules are currently on. A ligand binding flag can, if desired, be set in a similar fashion, based on the concentration of ligand and the K_d. Once the fast flags have been set, then the program inspects the reactions available to the two species A and B. The chemical change associated with each type of reaction (binding, phosphotransfer, methylation, etc.) is represented in the program together with "base values" of the reaction rate constants. The particular instantiation of the reaction, specified by the current state of the flags on A and B, is accessed from an array of values calculated at the beginning of the program, when the reaction system is being initialized. Values in the array modify the reaction probability according to the particular set of binary flags. In this manner, StochSim calculates a set of probabilities, corresponding to the reactions available to the particular states of molecules A and B, and then uses a random number to select which reaction (if any) will be executed in the next step. The reaction will be performed, if appropriate, and the relevant slow flag flipped.

Although it sounds complicated, the above sequence of events within an individual iteration takes place very quickly and even a relatively slow computer can carry out hundreds of thousands of iterations every second. Moreover, the strategy has the advantage of being intuitively simple and close to physical reality. For example, it is easy, if required, to label selected molecules and to follow their changes with time. Lastly, the speed of the program depends not on the number of reactions but on the numbers of molecules n in the reaction system (with a time of execution proportional to n^2). The StochSim distribution[2] consists of a platform-independent core simulation engine encapsulating the algorithm just described, together with separate graphical and user interfaces.

COMPARISON WITH THE GILLESPIE ALGORITHM

Daniel Gillespie showed, in the 1970s, that it is possible to simulate chemical reactions by an efficient stochastic algorithm (Gillespie, 1976). He showed that this algorithm gives the same results, on average, as conventional kinetic treatments (Gillespie, 1977), and later provided a rigorous mathematical derivation for the procedure (Gillespie, 1992). The Gillespie algorithm has since been used on numerous occasions to analyze biochemical kinetics, for example to simulate the stochastic events in lambda lysogeny (McAdams and Arkin, 1997; Arkin et al., 1998). In view of its evident success, the question therefore arises: Why in our work did we not use the Gillespie algorithm but chose to develop our own formulation? As shown in Appendix B, the Gillespie and StochSim algorithms are based

2 The latest version of StochSim can be obtained via FTP from `ftp://ftp.cds.caltech.edu/pub/dbray/`.

on equivalent fundamental physical assumptions. However, significant practical differences arise in applying the two algorithms to biochemical systems, as described below.

The Gillespie algorithm makes time steps of variable length, based on the reaction rate constants and population size of each chemical species. In each iteration, one random number is used to determine when the next reaction will occur, and another random number determines which reaction it will be. Both the time of the next reaction τ, and the type of the next reaction μ are determined by the rate constants of all reactions and the current numbers of their substrate molecules. Upon the execution of the selected reaction in each iteration, the chemical populations are altered according to the stoichiometry of the reaction, and the process is repeated. By avoiding the common simulation strategy of discretizing time into finite intervals, the Gillespie algorithm benefits from both efficiency and precision — no time is wasted on simulation iterations in which no reactions occur, and the treatment of time as a continuum allows the generation of an "exact" series of τ values based on rigorously derived probability density functions.

However, the efficiency of the Gillespie algorithm comes at a cost, and its precision is guaranteed only for chemical systems with certain properties. The efficient algorithm that selects which reaction to execute next and what time interval to take, does not represent each molecule in the system separately. With regard to the reactions of a typical cell signalling pathway, for example, it cannot associate physical quantities with each molecule, nor trace the fate of particular molecules over a period of time. Similarly, without the ability to associate positional and velocity information with each particle, the algorithm cannot be easily adapted to simulate diffusion, localization or spatial heterogeneity. Indeed, the "exactness" of the Gillespie algorithm holds only for spatially homogeneous, thermodynamically equilibrated systems in which non-reactive molecular encounters occur much more frequently than reactive ones (Gillespie, 1976, 1992).

A second limitation of the Gillespie algorithm (from a cell biological standpoint) is that it cannot easily handle the reactions of multistate molecules. Protein molecules are very frequently modified in the cell so as to alter their catalytic activity, binding affinity and so on. Cell signalling pathways, for example, carry information in the form of chemical changes such as phosphorylation or methylation, or as conformational states. A multi-protein complex may contain upwards of twenty sites, each of which can often be modified independently and each of which can, in principle, influence how the complex will participate in chemical reactions. With twenty sites, a complex can exist in a total of 2^{20}, or one million, unique states, each of which could react in a slightly different way. If our multi-protein complex interacts with only ten other chemical species, a detailed model may contain as many as ten million distinct chemical reactions, a combinatorial explosion. Any program in which the time taken increases in proportion to the number of reactions, as in a conventional, deterministic model, or in the Gillespie method, will come to a halt under these conditions.

We see therefore that STOCHSIM and the Gillespie algorithm take different approaches and are suited to different situations. STOCHSIM is likely to be slower than the Gillespie algorithm in calculating the eventual outcome of a small set of simple biochemical reactions, especially when the numbers of molecules is large. However, if the system contains molecules that can exist in a large number of states, then STOCHSIM will not only be faster but also closer to physical reality. It is easy, if required, to label selected molecules in this program and to follow their changes with time, including changes to their detailed post-translational modification and conformational states. Lastly, spatial structures can be incorporated into the STOCHSIM framework with relative ease, as one can directly define the spatial location of individual molecules — something that would be difficult to do with the Gillespie algorithm.

SPATIAL EXTENSIONS TO STOCHSIM

The original version of STOCHSIM (1.0) treated the entire reaction system as a uniformly mixed solution. Although this is clearly not how molecules are arranged within living cells, the omission of spatial heterogeneity has been a norm in biochemical simulations because it greatly facilitates modeling and reduces the computational load of simulation. However, as the resolution of our understanding of biochemical processes increases, it is becoming clear that even in bacteria, the spatial organization of molecules often play an important role (RayChaudhuri et al., 2001; Norris et al., 1996).

In the chemotaxis pathway, the membrane receptors are not only associated with the signalling molecules CheW and CheA in the cytoplasm, but also clustered together, usually at one pole of the cell (Maddock and Shapiro, 1993). The density of packing of molecules in the cluster implies a regular arrangement, and recent model building led to the proposal of a hexagonal lattice built from CheA and CheW into which receptor dimers are inserted in sets of three (Shimizu et al., 2000). An arrangement of this kind would create a "microenvironment" within the cytoplasm which could sequester certain molecules and exclude others simply through binding affinities (and without an internal membrane). A regular lattice would allow neighboring receptors to influence each other's activity by what has been termed "conformational spread" (Bray et al., 1998; Duke et al., 1999). We must also consider the time taken for diffusible components of the signal pathway, notably CheY and its phosphorylated derivative, which have to shuttle repeatedly between the receptor complex on the plasma membrane and the flagellar motors. Although this time is short, consistent with distances of less than a micrometer and a

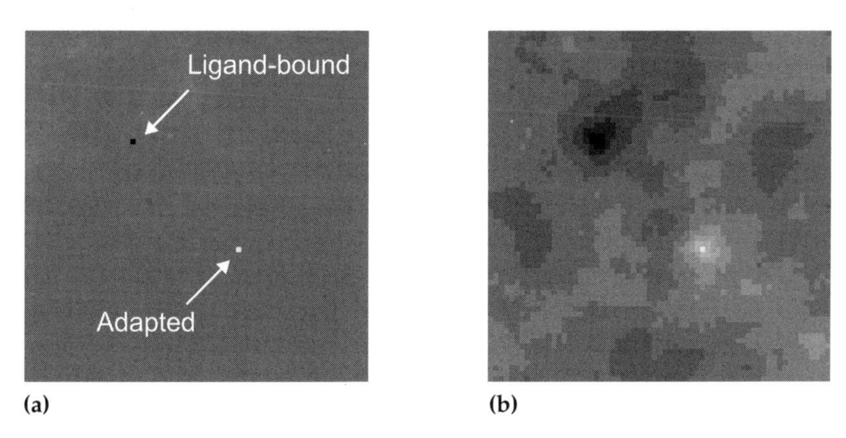

(a) (b)

Figure 10.1 Graphical representations of the spatial patterns of receptor activity in a STOCHSIM simulation of the *E. coli* chemotaxis pathway with clustered receptors. The two arrays shown above are both square lattices of 50 × 50 closely packed receptors, represented by one pixel each. These views are averaged over 0.1 seconds of simulated time, and the activities are represented by sixteen gray levels, with white corresponding to active and black to inactive receptors. In this simulation, ligand binding and methylation reactions were disabled in order to reveal the patterns due to activity spread alone. One receptor was permanently assigned to the ligand-bound (ligated and zero-methylated) state and another to the adapted (unligated and four-methylated) state. All other receptors were in the two-methylated state. In (a), no coupling reactions are defined and no spread of conformation is observed, but in (b), nearest-neighbor interactions allow the activity of certain receptors to "spread" over a wide range.

diffusion coefficient of around 5×10^{-8} cm^{2}s^{-1}, it is in principle measurable by recent techniques (Elowitz et al., 1999; Cluzel et al., 2000). A fully realistic model would have to deal with not only time delays but also the possibilities that diffusing species might have a non-uniform distribution and move within privileged channels in the interior of the cell.

Considerations such as these encouraged us to extend STOCHSIM to incorporate spatial representation, and the positions of the important molecular species within the cell. As a first step, we have so far introduced modifications that allow us to represent the two-dimensional arrangement of receptors in the plane of the plasma membrane. These changes, embodied in version 1.2 of STOCHSIM, assign two-dimensional coordinates to the array of receptors and permit such phenomena as the spread of activity from one receptor to its neighbor (Figure 10.1), and the diffusive movement of individual molecules bound to the surface array, to be represented.

Model of bacterial chemotaxis with "conformational spread"

In the *Escherichia coli* chemotaxis pathway, one of the major discrepancies between simulation and experiment thus far has been in the sensitivity of the system to very small changes in attractant concentration (Bray et al.,

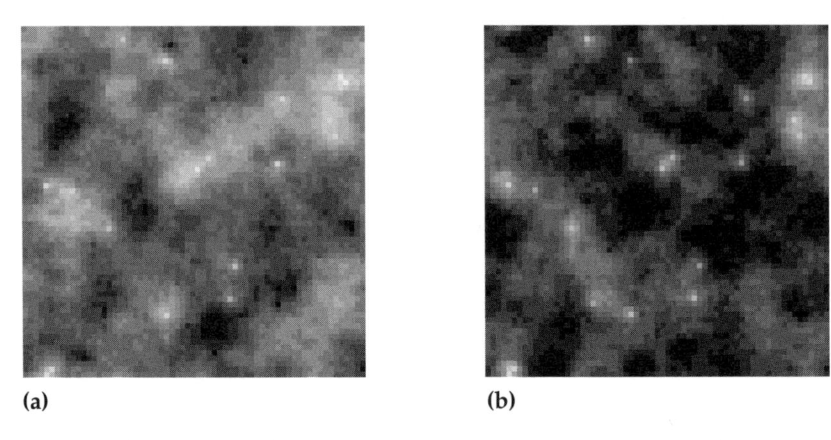

Figure 10.2 Patterns of receptor activity in an unconstrained simulation, in which ligand binding and methylation reactions were enabled in a coupled receptor array. These patterns are averaged over 0.01 seconds of simulated time, equivalent to roughly 16.7 million simulation iterations, in which each receptor flipped about 783 times on average. Discrete white patches correspond to the vicinities of adapted (highly methylated), and therefore highly active, receptors and black patches to centers of inactivity near bound ligand. In this simulation, the background attractant concentration was set to 10^{-8} M ($\sim \frac{1}{100} K_d$), and then doubled. The two patterns represent the average activity of intervals at 10 ms (a) prior to, and (b) after the doubling of stimulus. Rapid suppression of activity is observed, despite the very low concentration of stimulus.

1998). For example, computer-based estimates of the minimal detectable concentration of aspartate is on the order of 100 nM (Bray et al., 1998) whereas responses to concentration jumps as small as 5 nM have been detected experimentally (Segall et al., 1986). In order to test the possibility that the aforementioned "conformational spread" mechanism could account for this exquisite sensitivity observed in real bacteria, we have incorporated a two-dimensional representation of the receptor clusters into our previous STOCHSIM model (Morton-Firth et al., 1999) of the *E. coli* chemotaxis pathway. In that model, chemotactic receptors were modeled as multistate complexes with 11 binary flags to represent their various states. One of these flags represents the conformational state of the receptor, and was controlled by a rapid equilibrium. The probability of this flag being on or off, and hence the receptor being active or inactive, depended on the binding of ligand and the receptor's methylation state. In our new spatially extended model, an additional "coupling" factor that depends on the number of neighbors in the active state, has been defined. The more active neighbors a receptor has, the higher the probability of being active. The spread of conformations that results from this can be visualized in time-averaged views of the receptor cluster (Figures 10.1 and 10.2).

Preliminary results of our simulations indicate that the conformational spread mechanism can indeed serve to enhance the chemotactic response at the cost of higher steady-state noise. In simulations where receptor clus-

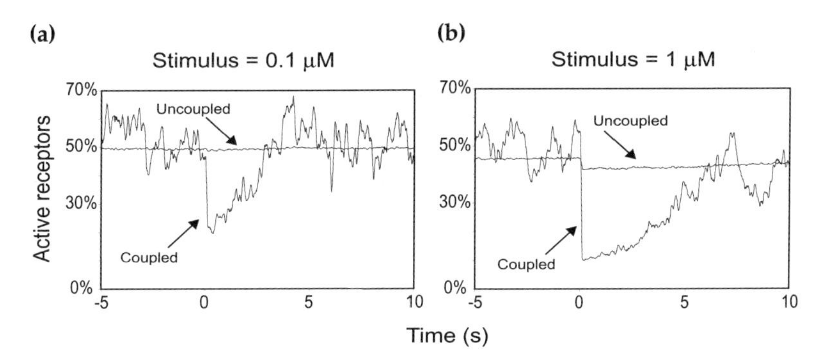

Figure 10.3 The enhancement of response achieved by coupling interactions in the receptor cluster model. Changes in total receptor activity during a doubling of stimulus at two background concentrations, (a) 0.1 μM and (b) 1 μM are shown. The concentration of ligand was doubled at time 0 in both (a) and (b). Significant enhancement is observed at both concentrations; the coupled array shows clear amplification of the ligand signal in (a), and in (b) only the coupled array shows a significant response to the doubling of attractant, demonstrating that the coupling could also act to increase the range of concentrations to which the system can respond.

ters were first adapted to various background concentrations of attractant and tested for their response to a subsequent doubling in stimulus, significant amplification of the signal is observed (Figure 10.3). The level of amplification is not as high as that reported previously (Bray et al., 1998; Duke et al., 1999), but the performance of the receptor cluster is less dependent on the precise value of the coupling strength. These differences arise because the previous models did not consider the multiple methylation states of the receptors, which can be covalently modified with up to four methyl groups. We are now investigating the effect of spatial patterns of methylation, an example of which is shown in Figure 10.4.

FUTURE DIRECTIONS

The obvious next steps for development of StochSim are the implementation of other geometries (e.g. triangular and hexagonal) for the two-dimensional arrays, further extending the spatial representation to a third dimension, and the development of a more generally accessible interface. Recent models of the neuromuscular junction include a realistic representation of the folds of the muscle membrane surface, the position and state of individual synaptic vesicles, and even the location of individual calcium ions (Stiles and Bartol, 2000). Three-dimensional representation of a cell may require some form of compartmentalization of its contents, whether into regularly spaced volume elements (voxels) or more biologically relevant compartments, such as nucleus, and membrane cortex. The interface development is now focused around a cross-platform

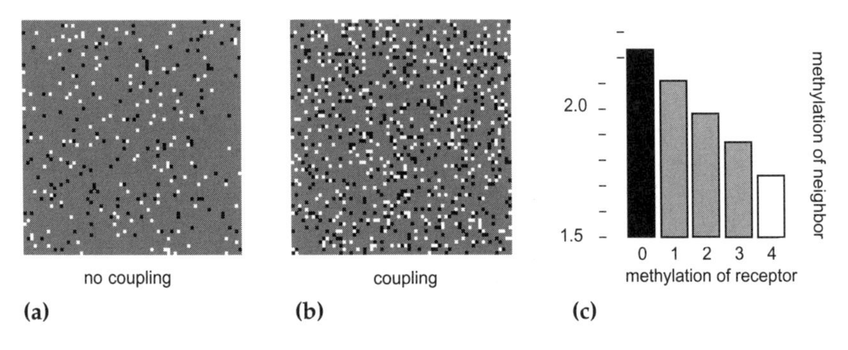

Figure 10.4 Spatial patterns of methylation. As in the activity snapshots (Figures 10.1 and 10.2), each pixel represents a single receptor in (a) an array without coupling and (b) an array with coupling. Four- and zero-methylated receptors (the extreme receptors) are are highlighted in black and white, respectively, while receptors in all other methylation states are shown in gray. A notable feature of the changes in methylation state distribution between the uncoupled and coupled arrays is that the extreme receptors are more abundant and tend to be closer together in the latter. This tendency is due to the relationship between receptor activity and methylation reactions, and is also reflected in (c) the average number of methyl groups per neighbor for each methylation state.

GUI (Le Novère and Shimizu, 2001), and a simple command-line interface is also provided for ease of scripting.

On the more general question of the future of modeling in cell biology, it seems unavoidable that stochastic representations will be increasingly useful. As the resolution of experimental techniques improves, they will generate large quantities of data relating to the behavior of individual cells and molecules. There will be an increasing emphasis on situations in which cell behavior depends on small numbers of molecules and analysis of such situations will naturally invoke individual-based simulations with a stochastic basis. Presently available simulation programs such as StochSim will doubtless become integrated into larger software packages that allow non-specialist users to quickly identify their requirements and obtain results. The remorseless increase in the power and speed of computers available to the modeling community will accompany, and empower, these developments.

ACKNOWLEDGMENTS

We wish to thank Matthew Levin for proofreading and providing critical comments on this manuscript, as well as Nicolas Le Novère, Tom Duke and Gavin Brown for helpful discussions.

APPENDIX A: DERIVATION OF REACTION PROBABILITIES IN STOCHSIM

First consider the following unimolecular reaction with substrate A:

$$\frac{d[A]}{dt} = -k_1[A] \tag{A.1}$$

If the size of the StochSim time-slice Δt is sufficiently small, the change in the number of reactant molecules, Δn_A, within this interval will be between 0 and 1, and is given by

$$\Delta n_A = -k_1 n_A \Delta t \tag{A.2}$$

where k_1 is the deterministic rate constant.

In the StochSim algorithm, the expected value of Δn_A within a single time-slice is

$$\begin{aligned} -\Delta n_A &= \text{Pr(molecule of A is selected in the first selection)} \\ &\quad \times \text{Pr(pseudo-molecule is selected in the second selection)} \\ &\quad \times p_1 \qquad \text{(A.3)} \\ -\Delta n_A &= \frac{n_A}{n} \times \frac{n_0}{n+n_0} \times p_1 \qquad \text{(A.4)} \end{aligned}$$

Equating Eqs. (A.2) and (A.4) gives

$$p_1 = \frac{k_1 n(n+n_0)\Delta t}{n_0}. \tag{A.5}$$

The probability for the bimolecular reaction can be derived similarly. Consider the following reaction with substrates B and C:

$$\frac{d[B]}{dt} = -k_2[B][C] \tag{A.6}$$

In a very small Δt,

$$\Delta n_B = -\frac{k_2 n_B n_C \Delta t}{2N_A V} \tag{A.7}$$

where V is the volume of the reaction system, and N_A is Avogadro's constant.

In StochSim, the expected value of Δn_B within a single time-slice is

$$\begin{aligned} -\Delta n_B &= \{\text{Pr(molecule of B is selected in the first selection)} \\ &\quad \times \text{Pr(molecule of C is selected in the second selection)} \\ &\quad \times p_2\} \\ &\quad + \{\text{Pr(molecule of C is selected in the first selection)} \\ &\quad \times \text{Pr(molecule of B is selected in the second selection)} \\ &\quad \times p_2\} \qquad \text{(A.8)} \\ -\Delta n_B &= 2 \times \frac{n_B}{n} \times \frac{n_C}{n+n_0} \times p_2 \qquad \text{(A.9)} \end{aligned}$$

Equating Eqs. (A.7) and (A.9) gives

$$p_1 = \frac{k_2 n(n + n_0)\Delta t}{2N_A V}. \tag{A.10}$$

APPENDIX B: EQUIVALENCE OF PHYSICAL ASSUMPTIONS IN THE GILLESPIE AND STOCHSIM ALGORITHMS

Gillespie rigorously derived his algorithm from what he called the *fundamental hypothesis* of stochastic chemical kinetics. To show that the StochSim algorithm can also be derived from this same hypothesis, we can translate the expressions for the StochSim probabilities (Eqs. 10.1 and 10.2) into the Gillespie formalism and show that in the limit $\Delta t \rightarrow 0$, it reduces to probability expressions derived directly from the *fundamental hypothesis*.

Gillespie's *fundamental hypothesis* states that the probability π of an elementary reaction R, occurring within the infinitesimal time interval δt, can be expressed as

$$\pi = hc\delta t \tag{B.1}$$

where h is the number of distinct molecular reactant combinations for the R reaction, and c is its stochastic rate constant. As with the deterministic rate constant k, the stochastic rate constant c can be interpreted to account for the mean rate at which reactant molecules collide, and the "activation energy" required for the reaction to occur. The relationship between the two constants are such that if the effects of fluctuations and correlations in reactant concentrations can be considered negligible, the following holds true for uni- and bi-molecular reactions (Gillespie, 1976):

$$k_1 \doteq c_1 \tag{B.2}$$

for the unimolecular rate constant k_1 and the unimolecular stochastic rate constant c_1 (both with dimensionality s^{-1}), and

$$k_2 \doteq N_A V c_2 \tag{B.3}$$

for the bimolecular rate constant k_2 (with dimensionality $M^{-1}s^{-1}$) and the bimolecular stochastic rate constant c_2 (with dimensionality s^{-1}). Here, V is the volume of the reaction system and N_A is Avogadro's constant.

For a specific unimolecular reaction R_1 with species A as reactant, h in Eq. (B.1) is simply the number of molecules of the reactant species ($h = n_A$). For a specific bimolecular reaction R_2 with species B and C as reactants, h is the product of the number of each species ($h = n_B n_C$). Therefore, the probability of an R_1 reaction occurring within δt is

$$\pi_1 = n_A c_1 \delta t, \tag{B.4}$$

and the probability of an R_2 reaction occurring within δt is

$$\pi_2 = n_B n_C c_2 \delta t. \tag{B.5}$$

We now proceed to inspect how Gillespie's reaction probabilities (π_1 and π_2) are related to the reaction probabilities in STOCHSIM. In doing so, it is important to note that the uni- and bi-molecular reaction probabilities stored in STOCHSIM's look-up tables (p_1 and p_2 in Eqs. 10.1 and 10.2) are conditional probabilities, i.e. they are the probability of a certain reaction occurring given that its reactant molecules have been chosen by STOCHSIM in the current time-slice. However, Gillespie's reaction probability π is the probability of a given reaction occurring within *any* given time interval, so we must first obtain the equivalent quantities for STOCHSIM reactions.

For a unimolecular reaction with a look-up table probability of p_1 in STOCHSIM, the probability ϖ_1 of this reaction occurring in any given time-slice can be written as

$$\varpi_1 = p_{uni} \cdot \frac{n_A}{n} \cdot p_1 \tag{B.6}$$

where $p_{uni} = n_0/(n + n_0)$ is the probability that a unimolecular reaction is tested for in each time-slice. Similarly for a bimolecular reaction whose look-up table probability is p_2,

$$\varpi_2 = p_{bi} \cdot 2 \cdot \frac{n_B}{n} \cdot \frac{n_C}{n} \cdot p_2 \tag{B.7}$$

where $p_{bi} = 1 - p_{uni} = n/(n + n_0)$ is the probability that a bimolecular reaction is tested for in each time-slice. Using p_{uni} and p_{bi}, Eqs. (10.1) and (10.2) can be rewritten as

$$p_1 = \frac{n k_1 \Delta t}{p_{uni}} \tag{B.8}$$

and

$$p_2 = \frac{n^2 k_2 \Delta t}{2 N_A V \cdot p_{bi}}. \tag{B.9}$$

Substituting (B.8) into (B.6)and (B.9) into (B.7) yields the following expressions for ϖ_1 and ϖ_2:

$$\varpi_1 = n_A k_1 \Delta t \tag{B.10}$$

$$\varpi_2 = n_B n_C \frac{k_2}{N_A V} \Delta t \tag{B.11}$$

If the assumptions made in obtaining Eqs. (B.2–B.3) are valid, we may further substitute (B.2) into (B.10) and (B.3) into (B.11), and take the limit $\Delta t \rightarrow 0$ to obtain

$$\varpi_1 = n_A c_1 \delta t = \pi_1 \tag{B.12}$$

and

$$\varpi_2 = n_B n_C c_2 \delta t = \pi_2. \tag{B.13}$$

We see, therefore, that the reaction probabilities for uni- and bi-molecular reactions calculated by the Gillespie and STOCHSIM algorithms are equivalent.

References

Arkin, A., Ross, J., and McAdams, H. H. (1998). Stochastic kinetic analysis of developmental pathway bifurcation in phage lambda-infected *Escherichia coli* cells. *Genetics* 149:1633–1648.

Bray, D. and Bourret, R. B. (1995). Computer analysis of the binding reactions leading to a transmembrane receptor-linked multiprotein complex involved in bacterial chemotaxis. *Mol. Biol. Cell* 6:1367–1380.

Bray, D., Bourret, R. B., and Simon, M. I. (1993). Computer simulation of the phosphorylation cascade controlling bacterial chemotaxis. *Mol. Biol. Cell* 4:469–482.

Bray, D., Levin, M. D., and Morton-Firth, C. J. (1998). Receptor clustering as a mechanism to control sensitivity. *Nature* 393:85–88.

Cluzel, P., Surette, M., and Leibler, S. (2000). An ultrasensitive bacterial motor revealed by monitoring signaling proteins in single cells. *Science* 287:1652–1655.

Duke, T. A. J., , and Bray, D. (1999). Heightened sensitivity of a lattice of membrane receptors. *Proc. Natl. Acad. Sci. USA* 96:10104–10108.

Ehlde, M. and Zacchi, G. (1995). MIST: a user-friendly metabolic simulator. *Comput. Appl. Biosci.* 11:201–207.

Elowitz, M. B., Surette, M. G., Wolf, P.-E., Stock, J. B., and Leibler, S. (1999). Protein mobility in the cytoplasm of *Escherichia coli*. *J. Bacteriol.* 181:197–203.

Gillespie, D. T. (1976). A general method for numerically simulating the stochastic time evolution of coupled chemical reactions. *J. Comput. Phys.* 22:403–434.

Gillespie, D. T. (1977). Exact stochastic simulation of coupled chemical reactions. *J. Phys. Chem.* 81:2340–2361.

Gillespie, D. T. (1992). A rigorous derivation of the chemical master equation. *Physica A* 188:404–425.

Goldbeter, A. (1996). *Biochemical Oscillations and Cellular Rhythms*. Cambridge University Press, Cambridge.

Hallett, M. B. (1989). The unpredictability of cellular behaviour: trivial or fundamental importance to cell biology? *Perspect. Biol. Med.* 33:110–119.

Hines, M. (1993). NEURON — a program for simulation of nerve equations. In *Neural Systems: Analysis and Modeling*, Eeckman, F. and Norwell, M. A., eds. Kluwer.

Karp, P. D., Riley, M., Saier, M., Paulsen, I. T., Paley, S. M., and Pellegrini-Toole, A. (2000). The EcoCyc and MetaCyc databases. *Nucleic Acids Res.* 28:56–59.

Kingston, R. and Green, M. R. (1994). Modeling eukaryotic transcriptional activation. *Curr. Biol.* 4:325–332.

Ko, M. S. H. (1991). A stochastic model for gene induction. *J. Theor. Biol.* 153:181–194.

Lamb, T. D. (1994). Stochastic simulation of activation in the G-protein cascade of phototransduction. *Biophys. J.* 67:1439–1454.

Le Novère, N. and Shimizu, T. S. (2001). StochSim: Modelling of stochastic biomolecular processes. *Bioinformatics* (in press).

Maddock, J. R. and Shapiro, L. (1993). Polar location of the chemoreceptor complex in the Escherichia coli cell. *Science* 259:1717–1723.

McAdams, H. H. and Arkin, A. (1997). Stochastic mechanisms in gene expression. *Proc. Natl. Acad. Sci. USA* 94:814–819.

McAdams, H. H. and Arkin, A. (1999). It's a noisy business! Genetic regulation at the nanomolar scale. *Trends. Genet.* 15:65–69.

Mendes, P. (1993). GEPASI: a software package for modelling the dynamics, steady states and control of biochemical and other systems. *Comput. Appl. Biosci.* 9:563–571.

Morton-Firth, C., Shimizu, T., and Bray, D. (1999). A free-energy-based stochastic simulation of the Tar receptor complex. *J. Mol. Biol.* 286:1059–1074.

Morton-Firth, C. J. (1998). Stochastic Simulation of Cell Signalling Pathways. Ph.D. thesis, University of Cambridge, Cambridge, UK CB2 3EJ.

Morton-Firth, C. J. and Bray, D. (1998). Predicting temporal fluctuations in an intracellular signalling pathway. *J. Theor. Biol.* 192:117–128.

Noble, D. (2001). *Oxford Cardiac Electrophysiology Group Website*. `http://noble.physiol.ox.ac.uk/`.

Norris, V., Turnock, G., and Sigee, D. (1996). The *Escherichia coli* enzoskeleton. *Mol. Microbiol.* 19:197–204.

RayChaudhuri, D., Gordon, G. S., and Wright, A. (2001). Protein acrobatics and bacterial cell polarity. *Proc. Natl. Acad. Sci. USA* 98:1332–1334.

Sauro, H. (1993). SCAMP: a general-purpose simulator and metabolic control analysis program. *Comput. Appl. Biosci.* 9:441–450.

Schaff, J., Fink, C. C., Slepchenko, B., Carson, J. H., and Loew, L. M. (1997). A general computational framework for modeling cellular structure and function. *Biophys. J.* 73:1135–1146.

Segall, J. E., Block, S. M., and Berg, H. C. (1986). Temporal comparisons in bacterial chemotaxis. *Proc. Natl. Acad. Sci. USA* 83:8987–8991.

Shimizu, T. S., Le Novère, N., Levin, M. D., Beavil, A. J., Sutton, B. J., and Bray, D. (2000). Molecular model of a lattice of signalling proteins involved in bacterial chemotaxis. *Nat. Cell Biol.* 2:792–796.

Smetters, D. K. and Zador, A. (1996). Synaptic transmission - noisy synapses and noisy neurons. *Curr. Biol.* 6:1217–1218.

Stein, L., Sternberg, P., Durbin, R., Thierry-Mieg, J., and J., S. (2001). WormBase: network access to the genome and biology of Caenorhabditis elegans. *Nucleic Acids Res.* 29:82–86.

Stiles, J. R. and Bartol, T. M. (2000). Monte Carlo methods for simulating realistic synaptic microphysiology using MCell. In *Computational Neuroscience*, De Schutter, E., ed. CRC Press, Boca Raton.

Stiles, J. R., Bartol, T. M., Salpeter, E. E., and Salpeter, M. M. (1998). Monte Carlo simulation of neurotransmitter release using MCell, a general simulator of cellular physiological processes. In *Computational Neuroscience*, Bower, J. M., ed. Plenum, New York.

Tjian, R. and Maniatis, T. (1994). Transcriptional activation: a complex puzzle with few easy pieces. *Cell* 77:5–8.

Tomita, M., Hashimoto, K., Takahashi, K., Shimizu, T. S., Matsuzaki, Y., Miyoshi, F., Saito, K., Tanida, S., Yugi, K., Venter, J. C., and Hutchison III, C. A. (1999). E-CELL: software environment for whole-cell simulation. *Bioinformatics* 15:72–84.

Van Steveninck, R. D. R. and Laughlin, S. B. (1996). Light adaptation and reliability in blowfly photoreceptors. *Int. J. Neural Systems* 7:437–444.

White, J. A., Rubinstein, J. T., and Kay, A. R. (1999). Channel noise in neurons. *Trends Neurosci.* 23:131–137.

Wilson, M. A., Bhalla, U. S., Uhley, J. D., and Bower, J. M. (1989). GENESIS: A system for simulating neural networks. In *Advances in Neural Information Processing Systems*, Touretzky, D., ed. Morgan Kaufmann, San Mateo, CA.

11 Computer Simulation of the Cell: Human Erythrocyte Model and its Application

Yoichi Nakayama and Masaru Tomita

We constructed a computer model of the human erythrocyte using E-CELL simulation system. The model has three major metabolic pathways including glycolysis, the pentose phosphate pathway, and nucleotide metabolism, as well as physical effects such as volume change along with osmotic pressure. In this paper, we report two results of simulation experiments as follows: (i) Analyses of the effect of osmotic pressure that changes the cell volume. (ii) The simulation of the hereditary enzyme deficiency of glucose-6-phosphate dehydrogenase (G6PD), including the pathways for glutathione (GSH) de novo synthesis and export system of glutathione disulfide (GSSG).

INTRODUCTION

In conventional molecular biology, experimental analyses of cells and organisms have been the primary focus. These experimental analyses have been performed within a limited system where the targeted phenomenon can be analyzed in a measurable form. Then, the acquired raw data of elementary processes are reorganized and reconstructed manually, in order to understand intracellular processes. Because of the fact that everything is reconstructed within the scope of human capacity, the scope for understanding the whole cell system is very limited. Thus, conventional molecular biology has not yet developed a capability of understanding and predicting the behavior of cells as a whole integrated system. In "Genome Project", however, which was initiated in the 80's, genomic sequences that cover all existing genomes have been collectively and all-inclusively identified. Genomic DNA was actually fragmented and the genomic sequence of each piece was determined by a system within the limited scope of conventional molecular biology. Information regarding an ever-expanding volume of genomic sequence, which is beyond human capacity to understand, has been computationally organized and integrated. New methods, such as Microarray analysis have continuously been developed for collective and high-throughput analyses of the intracellular events. "Cell Simulation" is, based on the expanded data acquired from such all-inclusive analyses and previously accumulated data, aims to analyze and understand the intact intracellular events by integrating all the information and reconstructing the various intracellular processes.

Comparable to the time when genomic sequences had been sporadically identified before "Genome Project" began, there have been so many worm-eaten holes in most parts of current data associated with intracellular processes. Stimulated by the great success of "Genome Project", however, projects that collectively analyze each intracellular process from various biological events have recently been initiated. Thus, in the near future, it will be possible to utilize the all-inclusive data of intracellular processes acquired from these projects. "Cell Simulation" will enable the reconstruction of cells in a computer using those data. "Cell Simulation" system has a great potential to become an important concept such as the weather forecast in experimental science.

We have been focusing on the development of "E-CELL", generic software for cell simulation, in our laboratory since 1996 (Tomita et al., 1999, 2000). In this article, we will review our goal and future aims. We will discuss several questions, such as what 'Cell Simulation' system aims at, and what will be possible through this 'Cell Simulation' system, by quoting our results of Cell simulation models reconstructed by the E-CELL system as an example.

SIMULATION ALGORITHM FOR CELLULAR PROCESSES

Cellular homeostasis is maintained by a myriad of different mechanisms. Major homeostatic mechanisms include "metabolism", "cell division" and "environmental adaptation". In order to simulate these cellular mechanisms, a certain operation where cellular process are abstracted and transformed into equations, is specifically required. This operation is termed "modeling". From a viewpoint of cellular simulation, it is very difficult to handle all the cellular processes within a single simulation algorithm. Instead, it requires the application of specific simulation algorithm most suitable for the properties of each cellular mechanism, by understanding each specific property. There are major two types of simulation algorithm currently used. One is "deterministic simulation", and another is "stochastic simulation".

"Deterministic simulation" can be calculated by phase equations, which express the equations of changes in several substances involved in certain cellular processes as a function of time. The calculated results of deterministic simulations are always identical how often the experiments are repeated, as far as the initial conditions are the same, since the calculation process for this deterministic simulation is completely independent of other indefinite factors such as probability. Therefore, deterministic simulation is very suitable for representing the average feature of certain events, such as some metabolic reactions, which occur simultaneously at many times. Since most experimental data can be obtained by analyzing a large group of cells considered to be almost identical, this method is generally most suitable when the calculation of simulation is started from the experimental values as an initial value. However, there are processes that may be very difficult to reconstruct by the deterministic simulation method. For example, in a certain regulation system easily affected by environmental changes, 50% of a group of cells exhibit one response and the remaining 50% of cells exhibit a completely different response. In this case, resultant events can be easily influenced by probabilities. If deterministic simulation is applied to represent this situation, all cellular processes are averaged out, resulting in a situation where all cells appear to exhibit identical reactions.

Stochastic simulation is the best solution for this problem. Two different stochastic simulation algorithms are used for the cell simulation, one of which determines reaction velocity by using probability, and another represents the situation by expressing the intermolecular interaction itself based on the probability. In both cases, the systems are based on the concept that intracellular processes are occurring under the control of probability. Therefore, it is possible to precisely express the "switch-like" stochastic process in the example described above. On the other hand, this method also has a defect. Since the reactions are expressed as a result of probability, multiple simulations are required when the average behavior of a certain group of cells should be obtained. It is up to a cellular process to determine exactly how many times simulations should be done in order to analyze the situation close to the average behavior of cells.

In addition to cases described above, there are several situations where diffusion and/or localization of certain molecular species need to be expressed as an intracellular process, and cases where the three-dimensional structure of macromolecules are required to be expressed.

SIMULATION OF METABOLIC PROCESSES

Many attempts have been made since the 1960's to reconstruct the intracellular metabolic pathways. One of the most basic concepts is the dynamic analysis of enzymatic reactions using rate equations. There were many simulation models of metabolic pathways published since then. The first practical cell simulation model was a human erythrocyte simulation model reported by Joshi and Palsson. This is a deterministic simulation model constructed by integrating models of the partial metabolic pathways, which had been previously reported and published. Their simulation model is reconstructed from major metabolic pathways of the erythrocyte, integrated with previously existing partial simulation systems of glycolysis, pentose phosphate pathway and nucleic acid metabolic pathway. The human erythrocyte has been well studied over the last three decades, and extensive biochemical data on its enzymes and metabolites have been accumulated. The erythrocyte of many species including the human erythrocyte does not contain a nucleus, nor does it carry genes. The cell uptakes glucose from plasma and processes it through glycolysis, generating ATP molecules for other cellular metabolism. The ATP molecules are consumed mostly for the ion transport systems in order to maintain electroneutrality and osmotic balance. Therefore, erythrocytes seem to be the best target for metabolic simulation due to their simplicity. However, the computer capacity at that time was very limited as compared to that we currently have. Therefore, it was almost impossible to carry out a long-term simulation for pathological analysis using this first erythrocyte simulation model. Various metabolic pathways including mitochondria, Calvin-Benson Cycle and glucose metabolism in pancreatic Langernhans islet cells had been modeled since then, but nothing led to a great success contributing to full understanding of the myths of life (biology). Meanwhile, generic simulation systems for the analysis of metabolic pathways such as KINSIM (Barshop et al., 1983), MetaModel (Cornish-Bowden and Hofmeyr, 1991), SCAMP (Sauro, 1993), MIST (Ehlde and Zacchi, 1995) and GEPASI (Mendes, 1993) have been developed. Moreover, generic simulation software including EX-TD and Stella are commercially available.

Our E-CELL project was launched in the 1990s. Using the E-CELL system, we developed a computer model of the human erythrocyte based on the previous model (Joshi and Palsson, 1989; Edwards and Palsson, 2000). The prototype model of a human erythrocyte consisted of glycolysis, pentose phosphate pathway, nucleotide metabolism and simple membrane transport systems. During calculation of steady state, we improved the parameters and kinetic equations based on experimental data in literature (Schauer et al., 1981; Schuster et al., 1989; Mulquiney and Kuchel, 1997). After repeating simulations with the addition of substance parameter estimations, the model has reached steady state, indicating that it is very close to approximating the real erythrocyte (Table 11.1).

Table 11.1 Steady state of the erythrocyte model. We obtained a data set of a steady state with this model. The initial data set was from experimental data in literature and predictions of previous simulation models. The simulation was run for more than 200,000 seconds in simulation time, until the model reached steady state.

Metabolic intermediate	ID	Initial value (mM)	Steady state (mM)	Experimental data (mM)
1,3-Diphosphogrycerate	13DPG	4.00E-04	1.83E-04	4.00E-04
2-Phosphogrycerate	2PG	1.40E-02	4.16E-03	1.40E-02 ± 5.00E-03
3-Phosphogrycerate	3PG	4.50E-02	4.62E-02	4.50E-02
Adenosine	ADO	1.20E-03	8.93E-06	1.20E-03 ± 3.00E-04
Dihydroxy acetone phosphate	DHAP	1.40E-01	1.35E-01	1.40E-01 ± 8.00E-02
Erythrose 4-phosphate	E4P	4.70E-04	1.17E+00	-
Fructose 6-phosphate	F6P	1.60E-02	6.39E-02	1.60E-02 ± 3.00E-03
Fructose 1,6-diphosphate	FDP	7.60E-03	1.14E-02	7.60E-03 ± 4.00E-03
Glucose 6-phosphate	G6P	3.80E-02	1.96E-01	3.80E-02 ± 1.20E-02
Glyceraldehyde 3-phosphate	GA3P	6.70E-03	6.24E-03	6.70E-03 ± 1.00E-03
Gluconolactone 6-phosphate	GL6P	1.17E-05	7.62E-06	-
Gluconate 6-phosohate	GO6P	1.86E-01	2.72E+00	-
Glutathione	GSH	3.21E+00	3.21E+00	3.21E+00 ± 1.50E+00
Glutathione	GSSG	1.06E-04	1.03E-04	-
Hypoxanthine	HXi	2.00E-03	9.32E-06	2.00E-03
Inosine monophosphate	IMP	1.00E-02	5.03E-03	1.00E-02
Inosine	INO	1.00E-03	3.32E-08	1.00E-03
Potassium	Ki	1.35E+02	1.26E+02	1.35E+02 ± 1.00E+01
Lactate	LACi	1.10E+00	1.20E+00	1.10E+00 ± 5.00E-01
Nicotinamide adenine dinucleotide	NAD	6.20E-02	8.87E-02	-
Nicotinamide adenine dinucleotide	NADH	2.70E-02	3.13E-04	-
Nicotinamide adenine phosphate	NADP	9.60E-05	8.06E-05	-
Nicotinamide adenine phosphate	NADPH	6.58E-02	6.58E-02	6.58E-02
Sodium	Nai	1.00E+01	2.27E+01	1.00E+01 ± 6.00E+00
Phosphoenolpyruvate	PEP	1.70E-02	1.89E-02	1.70E-02 ± 2.00E-03
5-Phosphoribosyl 1-phosphate	PRPP	5.00E-03	6.91E-05	5.00E-03 ± 1.00E-03
Pyruvate	PYRi	7.70E-02	6.00E-02	7.70E-02 ± 5.00E-02
Inorganic phosphate	Pi	1.00E+00	1.30E-01	1.00E+00
Ribose 1-phosphate	R1P	6.00E-02	2.12E-05	6.00E-02
Ribose 5-phosphate	R5P	3.30E-02	2.81E-04	-
Ribulose 5-phosphate	RU5P	1.29E-02	1.48E-04	-
Sedoheptulose 7-phosphate	S7P	2.30E-01	7.49E-02	-
Xylulose 5-phosphate	X5P	3.90E-02	4.30E-04	-
2,3-Diphosphogrycerate	2,3-DPG	4.50E+00	4.21E+00	4.50E+00 ± 5.00E-01
Adenosine diphosphate	ADP	2.70E-01	2.20E-01	2.70E-01 ± 1.20E-01
Adenosine monophosphate	AMP	8.00E-02	2.42E-02	8.00E-02 ± 9.00E-03
Adenosine triphosphate	ATP	1.54E+00	1.57E+00	1.54E-00 ± 2.50E-01

Stability of a simulation model: Effect of osmotic pressure on cell volume and metabolism

One of the very important factors is the stability and robustness of a system on the cellular simulation. Stability in this case means that the model of the entire metabolic system can accept whatever level of received changes, and can keep the conditions at a similar level prior to any changes. All living organisms create this "stability" using very subtle mechanisms in order to maintain homeostasis. In the field of engineering, there are two systems used for the improvement of stability of the systems; feed-back mechanism and redundancy. Feed-back system surely contributes to the stability of the cellular system. How about redundancy in the world of biology? This redundancy is very prominent specifically in an artificially constructed machine such as an airplane. This is because we can maintain a functionally high reliability of an individual machine. In biology, however, the reliability of one organism cannot be considered to be such a big factor. The organism, which is in failure of its original function after being mutated, should be useless, and could be beneficial when it is eliminated from the group where it belongs. In other words, the reliability of organisms is not important, but the maintenance of the species and adaptation to the new environment is crucial. Living organisms can survive even without the proper function of whichever pathway that has become redundant because of a certain cause. This results in the accumulation of mutants, which in turn leads to the acquisition of a new reason for the existence of one pathway (in most cases, another function) or selective elimination of another. What are the possibilities of a redundant pathway surviving? It is possible only if the reliability of individual functions is too low to threaten the existence of species, if it occurs with only one of the pathway with low reliability despite its great importance, or when one pathway becomes accidentally and temporarily redundant in the process of evolution.

The model, which we constructed, has been extended in various ways since then, and the second version is capable of simulating osmotic balance (Figure 11.1). The cell has to vary its volume while balancing osmotic pressure. In this model, the cell volume is made to increase or decrease until both osmotic pressures became equal. After this improvement, we analyzed the effect of this variable volume on metabolism. Metabolites of the cell at a steady state were increased/decreased, and the influence was observed. The graphs in Figure 11.2 show the differences between the fixed volume model and the variable volume model. In the variable volume model, the change was absorbed more quickly in all ten substances that we tried. These differences indicated a possibility that the volume, which changes with osmotic pressure, is stabilizing the metabolism. A possible mechanism of this stabilization is as follows: the cell volume changes with an increase or a decrease in a substance, and the concentration of all substances decreases or increases due to the change of cell volume. Then, enzymes, especially rate determining enzymes will alter their activity by their allosteric effect.

Furthermore, the reaction rates, which were oscillating in the fixed volume, were stabilized in the variable volume model (Figure 11.3). In this case, the amount of many substances will probably be changed synchronously, if only one reaction causes synchronous oscillation of many reactions. Therefore, the oscillation of reaction will be suppressed by the oscillation of cell volume.

We focused our effort mabinly on physical functions because the ki-

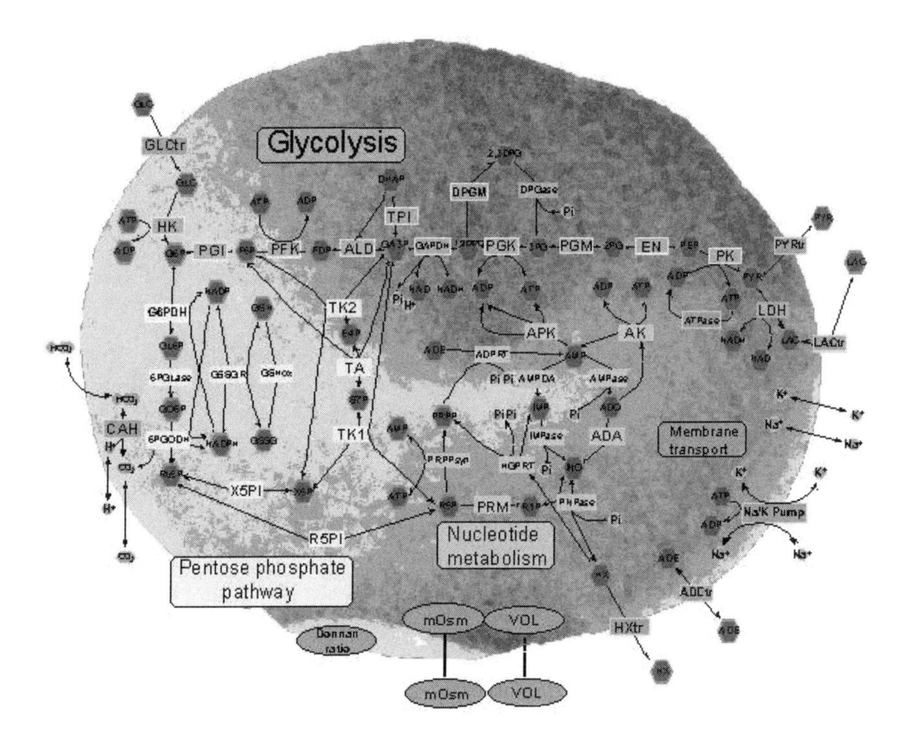

Figure 11.1 General concept and architecture of the erythrocyte model. The circles and hexagons are metabolic intermediates and ions. These molecular species are defined as "Substance" in the rule file of this model. Boxes represent enzymes and reaction processes. Their rate expressions are defined as "Reactor" whereas enzyme molecules are "Substance".

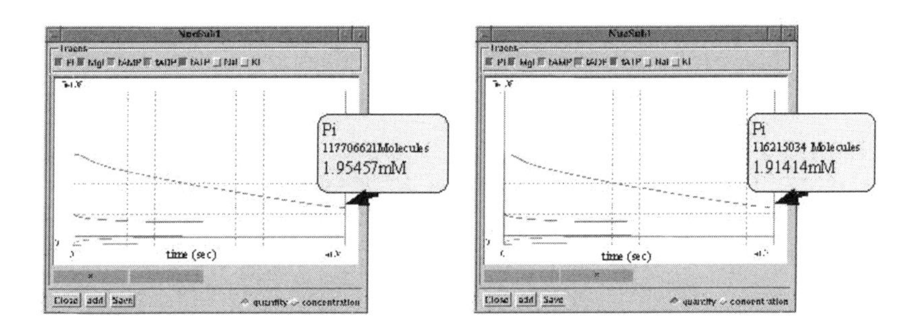

Figure 11.2 Effect of variable volume on metabolism. After starting the simulation, the number of molecules of phosphate was increased up to 50-fold at 100 sec, and the simulation was run for 30,000 seconds. The concentration of phosphate after 30,000-second simulation was about 1.91 mM in the variable volume model while it was about 1.95 mM in the fixed volume model.

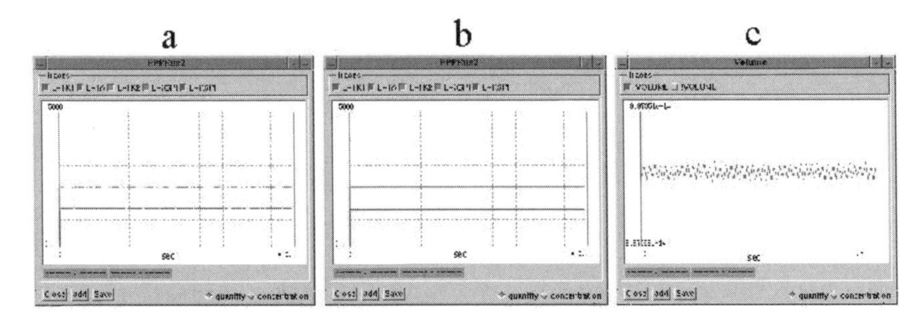

Figure 11.3 Effect of variable volume on reaction rates. (a) Reaction rates in the fixed volume model. (b) Reaction rates in the variable volume model. (c) Cell volume in the variable volume.

netic equations of various enzymes especially the rate-determining enzymes already contain the feedback factor. As already shown, the variable volume system can probably stabilize metabolism in the human erythrocyte. Similarly, it is plausible that pH have a role for stabilizing metabolism. The metabolic enzymes of red blood cells are affected strongly by the pH. Although the cell has buffering ability, the intracellular pH is influenced strongly by the pH of plasma, probably receiving non-negligible influences from substance amounts in cells. The accumulated or decreased amount of charged molecules in abnormal conditions should affect intracellular pH, and the pH sensitivity of enzymes is considered to be acting as a specific feedback system. As discussed above, in addition to the allosteric effects of enzymes, there seems to exist various physical feedback systems in cells. In the simulation, not only the rate equations of enzymes, but also these physical functions that occur in nature, must be modeled to achieve an accurate simulation model.

Simulation of G6PD deficiency and homeostasis

We next carried out the simulation of Glucose-6-phosphate dehydrogenase (G6PD). G6PD is a key enzyme that produces NADPH in the pentose phosphate pathway. G6PD converts glucose-6-phosphoric acid into 6-phosphoglucono-1,5-lactone, and generates NADPH simultaneously. After that, this metabolic intermediate is metabolized into ribulose-5-phosphoric acid via 6-phosphogluconic acid. In this process, NADPH is also generated. This reduction power is applied at various other intracellular processes, most importantly the reduction of GSSG. A major function of GSH in the erythrocyte is to eliminate superoxide anions and organic hydroperoxides. Peroxides are eliminated through the action of glutathione peroxidase, yielding GSSG.

At first, we implemented the inhibition of rate-determining enzymes caused by the ratio of GSH/GSSG to the model (Sugiyama, 1988), and simply modified the kinetic parameters of G6PD into those of the mutant extracted from patients with the deficiency (Jacobash et al., 1987). The simulation experiments were carried out with steady state concentrations corresponding to those of the normal erythrocyte. Sequential changes in the quantity of NADPH, GSH, and ATP were observed in the simulation experiments. However, the longevity of our computer model estimated by the concentration of ATP turned out to be much shorter than that of the real erythrocyte with G6PD deficiency. This difference was presumably due to the lack of pathways producing GSH and the export system of GSSG. A mature erythrocyte contains 2mM of GSH, but contains only several μM of GSSG. It is well known that in order to keep the ratio of GSH/GSSG stable, in addition to reduction with NADPH, other processes such as these pathways play an important part. We obtained the kinetic equations and parameters of these pathways from experimental analyses (Rae et al., 1990; Sriram and Ali-Osman, 1993), and implemented these pathways to our model. After the modification, the longevity of the cell was longer and the ratio of GSH/GSSG was higher as shown in Figure 11.4. This result indicates that these pathways compensate the reduction of GSH partially, and have a role to ease anemia, a condition of G6PD deficiency. This result can be a good explanation for the fact that G6PD deficiency is the most common cause of anemia. From the standpoint of evolution, if the deficiency has no severe disadvantage for surviving because of these compensation pathways, it would spread.

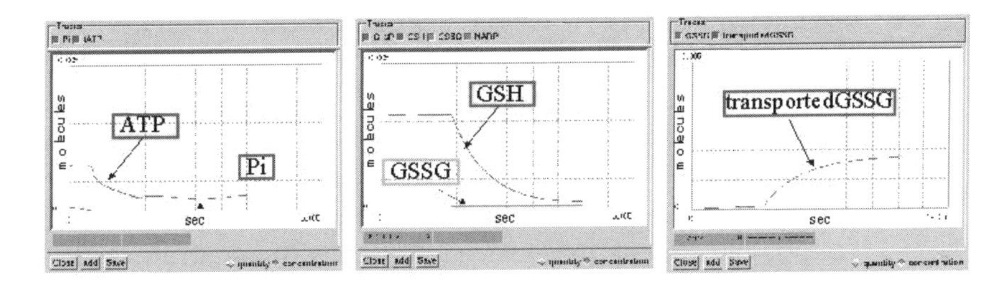

Figure 11.4 Simulation of G6PD deficiency. When the activity of G6PD is decreased, the activity of 6-phosphogluconate dehydrogenase increased, compensating the reduced production of NADPH. However, because 6-phosphoglucono-1,5-lactone was not supplied due or due to G6PD deficiency, in a short time, 6-phosphogluconic acid was exhausted and the production of NADPH was stopped. Then, the amount of NADPH started to reduce gradually and was soon exhausted, and reducing glutathione (GSH) started to decrease and was soon completely converted into GSSG. Then, the cells degenerated in their metabolic performance and finally exhausted all ATP because of the inhibition of rate-determining enzymes by low GSH/GSSG.

During this simulation analysis, we realized that the longevity of enzymes should be considered to calculate the concentration of metabolic intermediates at a steady state. For the mature erythrocyte, the activity of enzymes decreases gradually with time. We calculated the point with which production and consumption of all metabolic intermediates become equal as a steady state. However, this mathematical steady state does not represent the real steady state. Biological homeostasis is essentially different from "mathematical steady state". Such a condition never occurs in living organisms, especially in higher multi-cellular organisms. It can be said that the constancy of "homeostasis" in multi-cellular organisms is maintained by replacing the loss with disposable cells for a long time. It is speculated that these disposable cells never reach a mathematical steady state. A model, which can tolerate a long-term simulation for practical application to pathological analysis of human diseases, should not approximate to the "mathematical steady state". Moreover, in cases where the system reaches a steady state with a certain oscillation, it is impossible to obtain the biological homeostasis in actual living organisms. To solve this problem, we are trying to develop a method of searching for a real steady state using a parameter estimation method called the genetic algorithm.

SIMULATION OF THE SIGNAL TRANSDUCTION PATHWAY AND REGULATION NETWORK OF GENE EXPRESSION

Major simulation models previously constructed for the signal transduction are pathways associated with cell cycles (Aguda, 1999; Chen et al., 2000), bacterial chemotaxis (Bray et al., 1993), and circadian rhythm (Leloup and Goldbeter, 2000). These models mainly take deterministic simulations, but some models using stochastic algorithm have been constructed recently. The pilot study of stochastic simulation of the signal transduction pathways has been reported by Morton-Firth, Bray et al, who have constructed a model of the bacterial chemotaxis using the StochSim (Morton-Firth and Bray, 1998), a newly designed software platform for stochastic simulations. In this simulation model, each molecule is randomly selected and reacts with other selected molecules by the probability calculated from free energy. This chemotaxis simulation can very precisely express the function of the "Tar" receptor, and has successfully represented the mechanism "adaptation", which enables the sensing of subtle change (Morton-Firth et al., 1999). We are also simulating the bacterial chemotaxis. This simulation model is aimed at combining two algorithms, deterministic and stochastic simulations, within a single cell model. The motor-driven behavior and signal transduction pathways are expressed with stochastic and deterministic simulations respectively. In a whole cell simulation model, the development of a combined model is considered essential for the goal of accurate expression of integrated cellular processes, and for the reduction of calculation cost. A simulation model of lambda by McAdams and Shapiro is representative of simulation for the regulation network in gene expression. In addition to this, McAdams and Arkin have attempted to construct the simulation based on stochastic theory similar to that of the signal transduction pathways.

SIMULATION OF WHOLE CELL

A self-sustaining model is the first model that was constructed using our E-CELL system (Tomita et al., 1999, 2000). In this model which was reconstructed from genetic information of Mycoplasma Genitalium, the cells can uptake glucose from the outside to the inside of the cells by enzymes encoded by certain specific genes, and through the gene expression system within the cells, they metabolize the ingested glucose to produce ATP as an energy source (Figure 11.5). On the computer screen, gene-knock out can be easily done with the gene map window prepared in the E-CELL system. Moreover, it is possible to confirm the actual activities of metabolic enzymes and transcription and translation activities by the reactor window. Furthermore, the number of molecule for each substance can be confirmed by the substance window.

CONCLUDING REMARKS

It is nearly impossible to obtain a complete collection of accurate data to reconstruct cells in the computer. Recently, network prediction for gene regulation, signal transduction and metabolism has been highly valued as a basic technique for reconstruction of simulation models. However, it is theoretically very difficult to predict the precise networks from time series data. The reason for this is because the number of combination could be explosively large for the prediction of the networks except in cases that occur in a relatively closed system or in cases where most parts are already known. In the near future, however, it may be possible to directly reconstruct the networks from data related to gene-protein interaction and their roles obtained through protein-chips, two-hybrid methods and thorough analysis of gene-knock-out strains. Practically, however, there is also a major problem of the explosive number of combination of factors. For example, it is speculated that there are approximately 30,000 genes in humans. When attempting to determine whether all these genes bind to each other one to one (1:1), it is speculated that the number of the possible combinations reaches almost 900 million. Moreover, this number is the calculated result for only one to one binding, and it has been known that many proteins interact with two or more proteins. Thus, if you try to thoroughly analyze everything including the cases when several molecules form complexes, and so on, the number of possible combinations is explosively high. Furthermore, there are many false positives and false negatives observed in the "Two-Hybrid" method. On the other hand, in the thoroughly collective analysis of gene-knock out strains, there remain some problems in the reliability of data, since only the temperature-sensitive mutation or the complementation test can be analyzed for the essential genes. Another technical problem in the analysis of gene knock out strains is that sometimes, there are no phenotypes appearing in many cases with mutants in

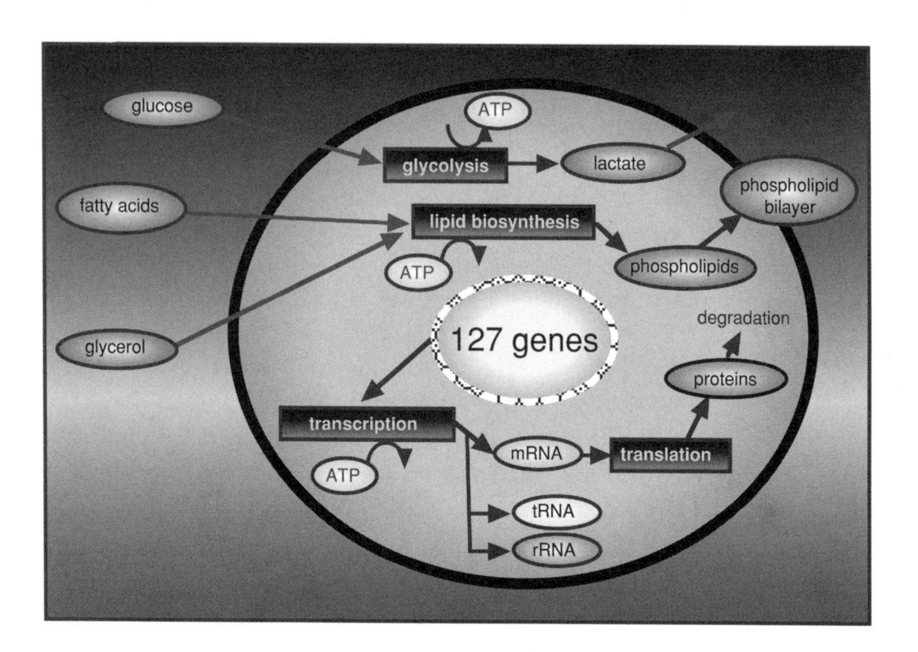

Figure 11.5 Overview of the cell model. We simulated the cell behavior in a condition where glucose is depleted using this model. As a result, the concentration of intracellular ATP transiently increases before it decreases. This is because ATP was no longer consumed at the earlier phase of glycolysis, but ATP was still produced by consuming the accumulated metabolic intermediates in the latter phase of glycolysis, resulting in a transient increase in ATP as a whole cell system. Such a result has never been observed or suggested in experiments. Thus, it has become possible to analyze and understand the behavior of the cells as a whole cell system using the cell simulation system integrated with various pathways. Examples of other successful models include the construction of cell simulation models by Goryanin et al and Schaff et al using DBSolve (Goryanin et al., 1999) and V-Cell (Schaff et al., 1997), respectively.

an experimental environment. Even in "Genome Project", however, the speed of progress in determination of gene sequences, which was thought impossible and adventurous when it began, have been conquered by various great progresses in high technology. Therefore, we believe that these practical problems in setting the experiments described above would be eventually solved in the near future. One may think that one of the solutions for this problem is probably the prediction of protein-ligand docking using amino acid sequences.

References

Aguda, B. D. (1999). A quantitative analysis of the kinetics of the G(2) DNA damage checkpoint system. *Proc. Natl. Acad. Sci. USA* 96:11352–11357.

Barshop, B.A., Wrenn, R. F., and Frieden, C (1983). Analysis of numerical methods for computer simulation of kinetic processes: development of KINSIM–a flexible, portable system. *Anal. Biochem.* 130:134–145.

Bray, D.; Bourret, R. B.; Simon, M. I. (1993). Computer simulation of the phosphorylation cascade controlling bacterial chemotaxis. *Mol. Biol. Cell* 4:469–482.

Chen, K. C.; Csikasz-Nagy, A., Gyorffy, B., Val, J., Novak, B., Tyson, J.J. (2000). Kinetic analysis of a molecular model of the budding yeast cell cycle. *Mol. Biol. Cell* 11:369–391.

Cornish-Bowden, A., and Hofmeyr, J.H. (1991). MetaModel: a program for modeling and control analysis of metabolic pathways on the IBM PC and compatibles. *Comput. Appl. Biosci.* 7:89–93.

Edwards, J.S., Palsson, B.O. (2000). Multiple steady states in kinetic models of red cell metabolism. *J. Theor. Biol.* 207:125–127.

Ehlde, M., and Zacchi, G. (1995). MIST: a user-friendly metabolic simulator. *Comput. Appl. Biosci.* 11:201–207.

Goryanin, I., Hodgman, T.C., and Selkov, (1999). Mathematical simulation and analysis of cellular metabolism and regulation. *Bioinformatics* 15:749–758.

Jacobash, G., Buckwitz, D., Jurowski, R., Gerth Ch., Plonka, A., and Kuckelkorn, U. (1987). Heterogeneity of Glucose-6-phosphate dehydrogenase enzymopathies in the GDR. *Biomed. Biochem. Acta.* 46:177–181.

Joshi, A. and Palsson, B.O. (1989). Metabolic Dynamics in the Human red Cell. Part I-A Comprehensive Kinetic Model. *J. Theor. Biol.* 141:515–528.

Ondo, T., Dale, G.L., and Beutler, E. (1980). Glutathione transport by inside-out vesicles from human erythrocyte. *Proc. Natl. Acad. Sci. USA* 77:6359–6362.

Leloup, J. C. and Goldbeter, A. (2000). Modeling the molecular regulatory mechanism of circadian rhythms in *Drosophila*. *Bioessays* 22:84–93.

McAdams, H.H. and Arkin, A. (1997). Stochastic mechanisms in gene expression. *Proc. Natl. Acad. Sci. USA* 94:814–819.

McAdams, H.H. and Shapiro, L. (1995). Circuit simulation of genetic networks. *Science* 269:650–656.

Mendes, P. (1993). GEPASI: a software package for modeling the dynamics, steady states and control of biochemical and other systems. *Comput. Appl. Biosci.* 9:563–571.

Morton-Firth, C.J., Bray, D. (1998). Predicting temporal fluctuations in an intracellular signalling pathway. *J. Theor. Biol.* 192:117–128.

Morton-Firth, C. J., Shimizu, T.S., and Bray, D. (1999). A free-energy-based stochastic simulation of the Tar receptor complex. *J. Mol. Biol.* 286:1059–1074.

Mulquiney, P.J. and Kuchel, P.W. (1997). Model of the pH-dependence of the concentrations of complexes involving metabolites, haemoglobin and magnesium ions in the human erythrocyte. *Eur. J. Biochem.* 245:71–83.

Rae, C., Berners-Price, S.J., Bulliman, B.T., and Kuchel, P.W. (1990). Kinetic analysis of the human erythrocyte glyoxalase system using 1H NMR and a computer model. *Eur. J. Biochem.* 193:83–90.

Sauro, H.M. (1993). SCAMP: a general-purpose simulator and metabolic control analysis program. *Comput. Appl. Biosci.* 9:441–450.

Schaff, J., Fink, C., Slepchenko, B., Carson, J., and Loew, L. (1997). A General Computational Framework for Modeling Cellular Structure and Function. *Biophys. J.* 73:1135–1146.

Schauer, M., Heinrich, R., and Rapoport, S.M. (1981). Mathematische Modellierung der Glykolyse und des Adeninnukleotidstoffwechsels menschlicher Erythrozyten. *Acta. biol. med. germ.* 40:1659–1682. (German)

Schuster, R., Jacobasch, G., and Holzhutter, H.G. (1989). Mathematical modelling of metabolic pathways affected by an enzyme deficiency. *Eur. J. Biochem.* 182:605–612.

Sriram, R. and Ali-Osman, F. (1993). Purification and Biochemical Characterization of gamma-Glutamylcysteine Synthetase from a Human Malignant Astrocytoma Cell line. *Biochem. Mol. Biol. Int.* 30:1053–1060.

Sugiyama, T. (1988). Modification of Enzyme Activity by Glutathione. *Tanpakushitsu Kakusan Kouso* 33:1423–1428. (Japanese)

Tomita, M., Hashimoto, K., Takahashi, K., Shimizu, T.S., Matsuzaki, Y., Miyoshi, F., Saito, K., Tanida, S., Yugi, K., Venter, J.C., and Hutchison, C.A. (1999). E-CELL: Software environment for whole cell simulation. *Bioinformatics* 15:72–84.

Tomita, M., Hashimoto, K., Takahashi, K., Matsuzaki, Y., Matsushima, R., Saito, K., Yugi, K., Miyoshi, F., Nakano, H., Tanida, S., Saito, Y., Kawase, A., Watanabe, N, Shimizu, T.S., and Nakayama, Y. (2000). The E-CELL Project: Towards Integrative Simulation of Cellular Processes. *New Generation Computing* 18:1–12.

Part V

System-Level Analysis

12 Constructing Mathematical Models of Biological Signal Transduction Pathways: An Analysis of Robustness

Tau-Mu Yi

Living organisms detect and respond to a variety of environmental cues (e.g., light, attractants, hormones, etc.) through receptor-mediated signal transduction pathways. The majority of pharmaceutical agents act by modulating the behavior of these biological networks. Many of the signaling systems possess remarkable performance characteristics including exquisite sensitivity, broad dynamic range, and robustness. Using techniques from control and dynamical systems theory, we have studied two well-characterized signaling pathways: (1) bacterial chemotaxis (two-component), and (2) mammalian visual phototransduction (G-protein). We have found that the dynamics of these complex networks are carefully regulated through feedback control in order to achieve robust performance.

INTRODUCTION

Understanding the robustness of biological systems is a major challenge facing biologists in the 21st Century. Robustness can be defined as the insensitivity of a particular system property to variations in the components and environment of the system. When designing and constructing a complex system such as a jet airplane, engineers are primarily concerned with robustness issues: can the plane fly in a wide range of weather conditions, can the plane fly if one or more instruments fail, etc.? Likewise, biologists who are trying to reverse engineer Nature's design should be equally concerned with robustness.

The main question that I am concerned with in this chapter is how biological systems achieve robust performance despite facing significant fluctuations in the external environment and the internal conditions. Changes in temperature, pH, nutrient levels, etc. contribute to the environmental uncertainty; mutations, variations in protein levels, aging, etc. contribute to the uncertainty of the components of living systems.

The short answer to this question is feedback control. By feedback control, I mean that you can regulate a certain property by measuring that property and feeding the information back into the system. For example, a thermostat measures the temperature of a room and based on that measurement decides whether to increase or decrease the heat.

I will focus primarily on a particular type of biological system: signal transduction networks. Living organisms detect and respond to a variety of environmental cues through signal transduction pathways. A typical signaling system consists of a receptor residing on the cell surface that binds a ligand representing the signal. The binding of ligand modulates the activity of the receptor which triggers a signaling cascade leading to an effector which produces a response. Each of these levels in the pathway are carefully regulated.

This chapter will be divided into two parts. For the first section, I will describe the robustness of a steady-state property, perfect adaptation in bacterial chemotaxis. In the second section, I will describe the robustness of a transient property, the single-photon response in phototransduction. In both cases, I will focus on specific feedback strategies used to ensure the robustness of each process.

ROBUST PERFECT ADAPTATION AND INTEGRAL FEEDBACK CONTROL IN BACTERIAL CHEMOTAXIS

In the bacterial chemotaxis signaling network, the receptor complex – which consists of receptor, the histidine kinase CheA, and the adaptor protein CheW – phosphorylates the response regulator CheY (Stock and Surette, 1996). Phosphorylated CheY, interacts with the flagellar motor to induce tumbling. Attractant inhibits the receptor complex producing straight runs. Receptor complex activity is regulated by methylation. Methylation by CheR increases activity. Demethylation by CheB decreases activity. In this work, I assume that CheB only demethylates active receptor complexes (Barkai and Leibler, 1997), thus allowing CheB to sense the activity state of the receptor. This assumption leads to a key negative feedback loop acting through CheB (see Figure 12.1).

Bacterial chemotaxis exhibits perfect adaptation. Experimental data (Berg and Brown, 1972; Macnab and Koshland, 1972) showed that a continual dose of attractant produced a transient increase in the output, followed by a period of adaptation, and then a return to the prestimulus level of activity, Y_0. Thus, the steady-state level of activity, Y_{SS}, asymptotically approached Y_0, and this was observed for a wide range of attractant concentrations. Thus, perfect adaptation can be defined as $Y_{SS} = Y_0$ for all concentrations of attractant.

Recently, Leibler and colleagues tested the robustness of perfect adapation to dramatic changes in the concentration of key components of this pathway (Alon et al., 1998). They demonstrated that as the methylase CheR was varied over a 50-fold range, the output Y_{SS} remained close to Y_0. They went on to show that perfect adaptation was also robust to

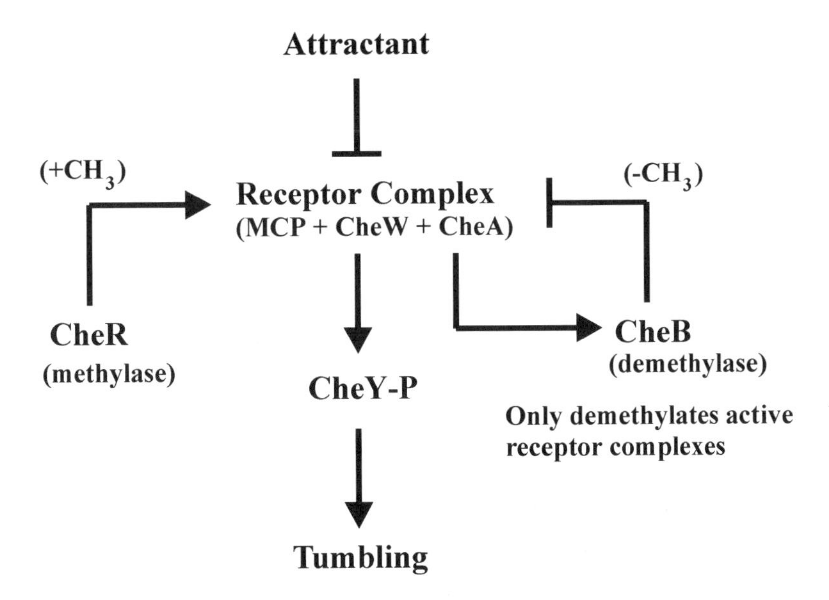

Figure 12.1 Schematic diagram of the bacterial chemotaxis signal transduction pathway.

changes in the levels of CheB, receptor, and CheY.

Is it possible to model perfect adaptation in bacterial chemotaxis? Several models in the literature indeed were able to reproduce perfect adaptation, but only through the fine-tuning of the model parameters (Hauri and Ross, 1995; Spiro et al., 1997). Perfect adaptation was non-robust in these models because subtly altering a parameter disrupted perfect adaptation. Alternatively, one can imagine that perfect adaptation is a structural property of the system insensitive to parameter variation, perhaps resulting from a particular feedback control mechanism. To distinguish between these two types of models, we have systematically varied model parameters and tested for perfect adaptation in two different models (Yi et al., 2000).

In one example, the model of chemotaxis developed by Spiro et al. (1997), the total concentration of receptor was varied over a 100-fold range from 1 μM to 100 μM, and the steady-state level of receptor complex activity was evaluated at three different levels of ligand concentration (0, 1 μM, 1 mM). Perfect adaptation occurred at only a single value of this parameter, 8 μM of total receptor concentration. In this model, perfect adaptation was non-robust because changing total receptor concentration from 8 to 10 μM did not preserve perfect adaptation.

The model of Barkai and Leibler was different (Barkai and Leibler, 1997). In their model, the steady-state receptor activity at the three concentrations of attractant completely superimposed as you varied the total receptor concentration. Perfect adaptation was robust to a 100-fold range in receptor concentration and to dramatic changes in the levels of the other protein components of this system as well as to perturbations in the kinetic rate constants.

The key to the Barkai-Leibler model is what they term activity dependent kinetics which leads to robust perfect adaptation. To control engineers this type of behavior is quite familiar; it is a sign of integral feedback control.

What is integral feedback control? It is a type of feedback structure that ensures the robust tracking of a specific steady-state value so that the error approaches 0 despite parameter variation. The term integral refers to the fact that the time integral of the system error is fed back into the system, not the error itself. Integral controllers are ubiquitous in man-made systems. For example, a thermostat uses integral control to maintain robustly the temperature in a room at the set point despite doors being opened and closed and despite the heater not performing up to specifications.

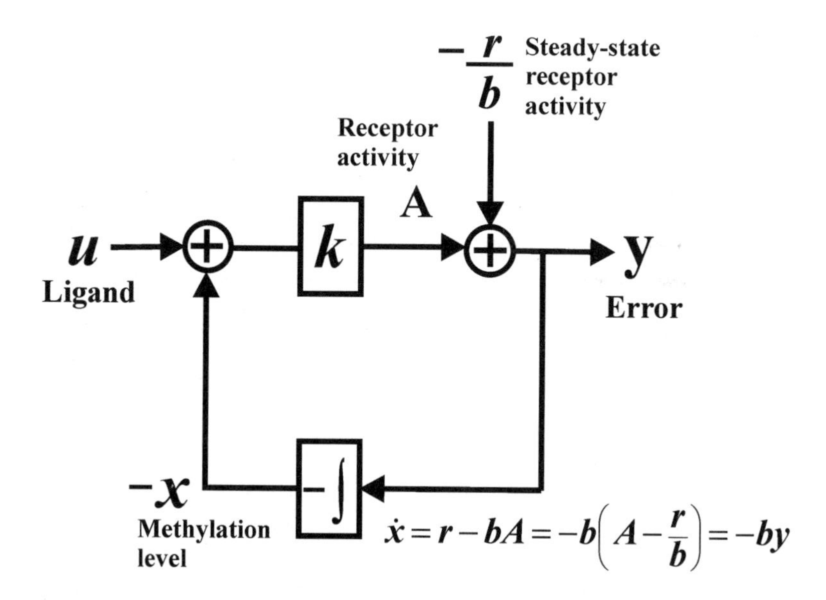

Figure 12.2 Block diagram of integral feedback control and bacterial chemotaxis. The input u is the ligand concentration, and the block with gain k represents the receptor dynamics producing the output A, receptor activity. The integral of the error y is the feedback term x which represents the methylation level of the receptor.

A block diagram of integral control illustrates its chief features (see Figure 12.2). The plant or network, schematically represented by the block with gain k, takes the inputs and produces the output A. The difference between the output A and the desired steady-state output is the error term y. This term is integrated and its negative is fed back into the system. The feedback term $x = \int y$, and so $\dot{x} = y$. At steady-state, $\dot{x}$ goes to 0, and hence y approaches 0 as t goes to infinity independent of the values of the input u and the gain k. Hence, the error asymptotically approaches 0 for all values of u and k as long as the system is stable.

We were able to derive the integral control equations from the Barkai-Leibler model thus demonstrating that their model possessed integral feedback (Yi et al., 2000). A simplified version of the derivation is shown here. The variable x represents the the total methylation state of the receptor. The change in x, $\dot{x}$, equals the methylation rate r minus the demethylation rate. Taking advantage of the assumption that CheB only demethylates active receptor complexes, we can write the demethylation rate as a function of A the receptor activity level. Thus, we obtain the following differential equation:

$$\dot{x} = r - bA \tag{12.1}$$

where bA is the activity-dependent demethylation rate. At steady-state, $r = bA$, and hence the steady-state receptor activity level is r/b. We can then rewrite equation (12.1) in terms of the error term y:

$$\dot{x} = r - bA = -b(A - \frac{r}{b}) = -by \tag{12.2}$$

Thus, we obtain the characteristic $\dot{x} = y$ equation for integral control.

We can superimpose the chemotaxis signaling network on the block diagram for integral control (see Figure 12.2). As we know the feedback loop arises through the methylation dynamics of the receptor, and x the feedback term approximates the methylation level of the receptor complex. The assumption that CheB only demethylates active receptor complexes leads to the characteristic equation for integral control. It all makes sense because the output A or receptor complex activity depends on the inputs u, ligand concentration, and x, methylation level.

Leibler and collegues essentially rediscovered integral control in the context of their model of bacterial chemotaxis. Our contribution was to place their findings within the framework of control theory. Another fundamental result of control theory states that integral control is not only sufficient for robust tracking but it is also necessary in linear systems. Thus, for the linearized approximation of the Barkai-Leibler model, we argue that integral control is the only feedback mechanism that can explain robust perfect adaptation. This suggests that even if the Barkai-Leibler model is later overturned, there is most likely some other mechanism for implementing integral feedback.

To a biologist this robust tracking of a set point sounds very familiar; it sounds like homeostasis. Homeostasis is the dynamic self-regulation observed in living organisms resulting in the maintenance of a relatively constant internal state (Fell, 1997). Thus, we suggest that integral control may represent an important strategy for ensuring homeostasis. For example, the intracellular concentration of calcium is influenced by numerous biological events. An integral control loop, perhaps acting through a regulatory enzyme like CaM-Kinase II which in turn may phosphorylate some calcium channel, could fix the steady-state concentration of intracellular calcium at some desired level. Variations in the calcium dynamics elsewhere in the cell would not affect this steady-state concentration.

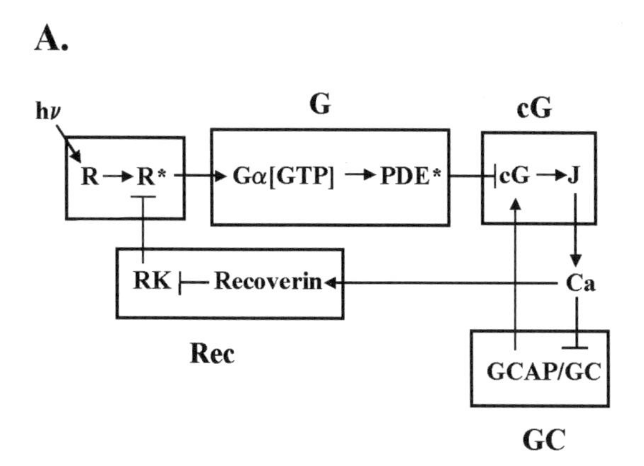

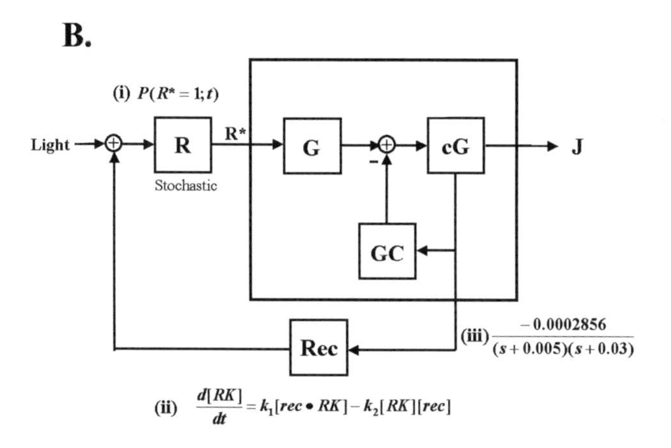

Figure 12.3 Phototransduction pathway. A. Schematic diagram of pathway divided into modules. B. Converting the schematic diagram of the pathway into a mathematical model. The arrows from the diagram above were replaced with mathematical equations.

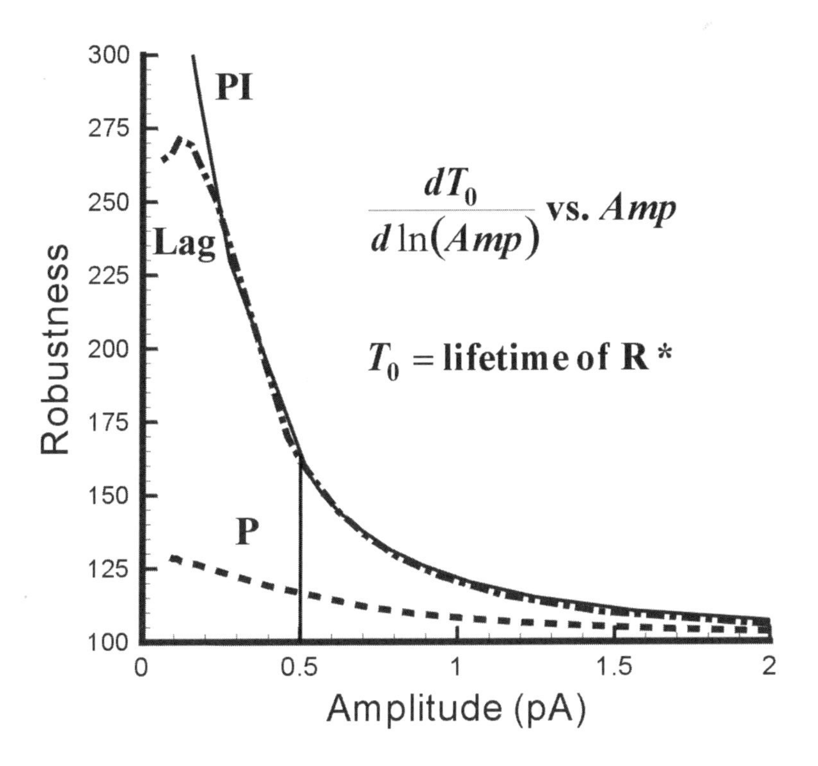

Figure 12.4 Robustness and feedback control in the single-photon response. The robustness of the amplitude of the current output to variations in T_0, the lifetime of R*, was measured for three controllers as a function of the amplitude. Robustness increased, and the amplitude decreased, as one raised the gain of the feedback. The curves for the proportional (P, dashed), proportional-integral (PI, solid), and lag (Lag, dashed-dot) controllers are shown.

REPRODUCIBILITY OF THE SINGLE-PHOTON RESPONSE AND CALCIUM-MEDIATED FEEDBACK IN PHOTOTRANSDUCTION

Let us now switch gears and turn our attention to the robustness of a transient process, the single-photon response in phototransduction. In the retina, rod photoreceptor cells detect light along with the cone photoreceptor cells and send this information to the visual cortex in the brain. The rod cell possesses a highly specialized geometry. In particular, the outer segment of the photoreceptor cell consists of many membraneous disks which are loaded with the components of the phototransduction signaling cascade.

When exposed to a single photon, rod photoreceptor cells produce a stereotyped current response. Early on, Baylor and colleagues noted the striking reproducibility of the single-photon response (Baylor et al., 1979). For example, Whitlock and Lamb (1999) recorded 50 trials from a single photoreceptor cell. The amplitude, time-to-peak, and decay constant from the resulting single-photon current responses were quite reproducible. Indeed, one consequence of this reproducibility is that one can easily distinguish the single-photon response from the zero-photon response from the two-photon response.

In Figure 12.3A, I schematically outline the phototransduction pathway. For simplicity, I have divided the pathway into modules. The receptor dynamics of the single-photon response consist of a single receptor activated by a photon and shut off by phosphorylation by rhodopsin kinase (RK) followed by binding to arrestin. The activated rhodopsin, R*, activates hundreds of G-proteins which in turn stimulate cGMP phosphodiesterase, PDE. Activated PDE, PDE*, degrades the second messenger cGMP, and this reduction in cGMP concentration results in the closing of cGMP-gated channels, causing a change in current. Note the two feedback loops acting via calcium. Calcium enters the cell through the cGMP channels, and hence the concentration of intracellular calcium is a function of the current J. In one loop, the enzyme guanylate cyclase (GC), which synthesizes cGMP, is modulated by the calcium-binding protein guanylate cyclase activating protein (GCAP). In the other loop, rhodopsin kinase is modulated by the calcium-binding protein recoverin (Stryer, 1991).

In my modeling I have preserved this basic structure and replaced the arrows with mathematical equations (see Figure 12.3B). I used several different types of equations. Because the receptor dynamics are stochastic, the level of R* at a given time t must be represented by a probability function (i). Hundreds of molecules of G-protein and cGMP phosphodiesterase are activated by the single R*. After this amplification, the concentration of the downstream species can be represented deterministically by ordinary differential equations, ODEs (ii). Finally, to simplify the analysis, I have approximated some ODEs by linear transfer functions (iii).

As described above, the single activated rhodopsin is phosphorylated by rhodopsin kinase, and the resulting phosphorylated molecule is bound by arrestin and eliminated. We can represent these stochastic events as a Poisson process. Furthermore, it is clear that these stochastic fluctuations in R* lifetime can contribute significantly to the variability in the single-photon response.

Let us now focus on the dynamics within the box (modules G, cG, and GC) depicted in Figure 12.3B. The input to the box is the number of activated rhodopsin molecules. The output is the current trace, J. To a control engineer, the problem is to design a controller that minimizes the variations in the output resulting from variations in the input and in the components of the system. Criteria for evaluating the potential controllers include the following: (1) robustness, (2) fit to real data, and (3) physical realizability.

We replaced the guanylate cyclase (GC) module with three types of controllers: (1) proportional controller (P), (2) proportional-integral controller (PI), and (3) lag compensator (Lag). The proportional controller takes the output and multiplies it by a constant before feeding it back into the system. The proportional-integral control combines a proportional controller with an integral controller. The lag compensator can be viewed roughly as a tunable PI controller.

How well do these three controllers suppress the variability in the response caused by the stochastic fluctuations in R* lifetime? I have plotted in Figure 12.4 the inverse sensitivity or robustness in the amplitude of the output to changes in the lifetime of R*, T_0. As one increases the gain of the feedback for all three controllers the robustness increases, but the amplitude decreases, resulting in a tradeoff. The wild-type response has an amplitude of 0.5 pA, and at that value the PI and lag controllers suppress variability in T_0 better than the proportional controller.

The closed-loop response of which controller can best fit data from rod photoreceptor cells derived from wild-type and mutant mice? Ideally, one would like to fit the response of the system to a step input. Knock-out mice have been created in which the rhodopsin kinase (RK) gene has been deleted (Chen et al., 1999). In these cells, R* is not inactivated for a long time, and after absorbing a single photon the resulting input to the system will be a step function.

In Figure 12.5, the black line represents the single-photon response from RK -/- rod cells. The output from the system using a proportional controller shows a slow rise. The PI controller response rises quickly but eventually returns to zero. Only the system with the lag compensator is able to provide an adequate fit to the real data.

Two populations of the GC/GCAP complex reside on the surface of the membraneous disks of the photoreceptor cell. One population faces the plasma membrane, the other is exposed to the interstitial region between disks. The concentration of calcium in the narrow region between the plasma membrane and the disk is expected to be proportional to the current J because of rapid equilibration; the concentration of calcium in the larger region between the disks is expected to be proportional to the integral of the current J over a defined period of time. Taken together, one would expect the calcium regulated activity of GC/GCAP to produce the combined proportional and integral action of a lag controller.

I have suggested that the calcium-mediated modulation of GC by GCAP may act as a lag compensator to enhance the robustness of the single-photon response. It is important to emphasize that other factors also contribute to the reproducibility of this response. For example, Rieke and Baylor (1998) have argued that the multi-step shutoff of activated rhodopsin serves to average the stochastic variability in R* lifetime. Alternatively, the calcium-mediated feedback loop acting through recoverin may modulate the behavior of R* as well.

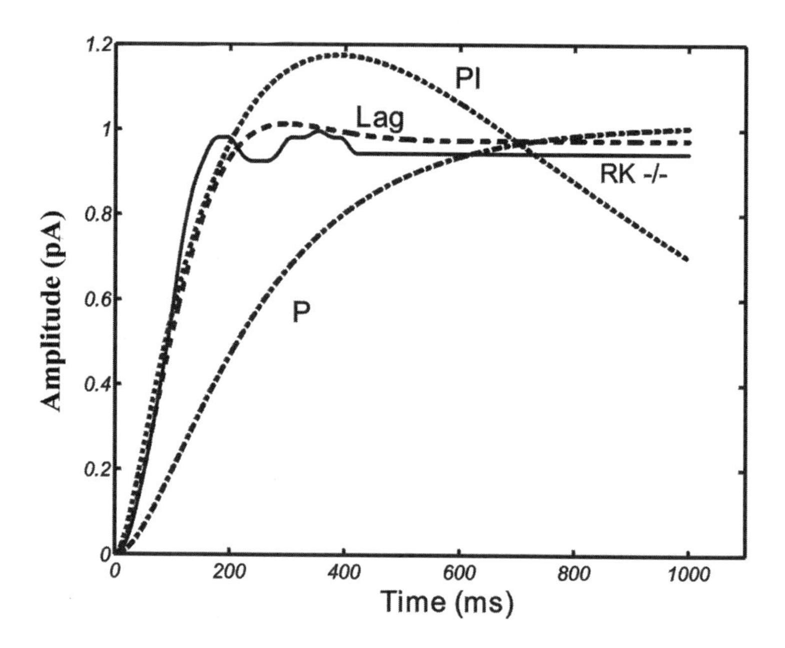

Figure 12.5 Fitting the closed-loop response of the three controllers to the single-photon response from RK -/- rod cells. Displayed are the experimental data (solid) as well as the closed-loop responses for the three controllers: (i) P (dashed-dot), (ii) PI (dotted), (iii) Lag (dashed).

CONCLUSIONS

One goal of this chapter is to convince the reader that robustness is an essential property of any biological system and that feedback control is a good way of achieving robustness. Thus, some of the most important features of any biological pathway are the feedback loops. The richness and complexity of biological processes arise through the dynamics of highly interconnected networks; feedback regulates the behavior of these networks. Indeed, control theory suggests that this dense web of interconnections may be necessary to achieve robust performance given the uncertainties faced by living systems.

References

Alon, U., Surette, M.G., Barkai, N., and Leibler, S. (1998). Robustness in bacterial chemotaxis. *Nature* 397:168–171.

Barkai, N. and Leibler, S. (1997). Robustness in simple biochemical networks. *Nature* 387:913–917.

Baylor, D.A., Lamb, T.D., and Yau, K.-W. (1979). Responses of retinal rods to single photons. *J. Physiol.* 288:613–634.

Berg, H.C. and Brown, D.A. (1972). Chemotaxis in *Escherichia coli* analysed by three-dimensional tracking. *Nature* 239:500–504.

Chen, C.-K., Burns, M.E., Spencer, M., Niemi, G.A., Chen, J., Hurley, J.B., Baylor, D.A. and Simon, M.I. (1999). Abnormal photoresponses and light-induced apoptosis in rods lacking rhodopsin kinase. *Proc. Natl. Acad. Sci. USA* 96:3718–3722.

Fell, D. *Understanding the Control of Metabolism* (1997). London: Portland Press.

Hauri, D.C. and Ross, J. (1995). A model of excitation and adaptation in bacterial chemotaxis. *Biophys. J.* 68:708–722.

Macnab, R.M. and Koshland, D.E., Jr. (1972). The gradient-sensing mechanism in bacterial chemotaxis. *Proc. Natl. Acad. Sci. USA* 69:2509–2512.

Rieke, F. and Baylor, D.A. (1998). Origin of reproducibility in the responses of retinal rods to single photons. *Biophys. J.* 75:1836–1857.

Spiro, P.A., Parkinson, J.S., and Othmer, H.G. (1997). A model of excitation and adaptation in bacterial chemotaxis. *Proc. Natl. Acad. Sci. USA* 94:7263–7268.

Stock, J.B. and Surette, M.G. (1996). In *Escherichia coli and Salmonella: Cellular and Molecular Biology*, ed. Neidhardt, F.C. ASM Press, Washington. pp. 1103-1129.

Stryer, L. (1991). Visual excitation and recovery. *J. Biol. Chem.* 266:10711–10714.

Whitlock, G.G. and Lamb, T.D. (1999). Variability in the time course of single photon responses from toad rods: termination of rhodopsin's activity. *Neuron* 23:337–351.

Yi, T.-M., Huang, Y., Simon, M.I., and Doyle, J. (2000). Robust perfect adaptation in bacterial chemotaxis through integral feedback control. *Proc. Natl. Acad. Sci. USA* 97:4649–4653.

13 Combination of Biphasic Response Regulation and Positive Feedback as a General Regulatory Mechanism in Homeostasis and Signal Transduction

Andre Levchenko, Jehoshua Bruck, and Paul W. Sternberg

Q: How is this type of regulation distinguishable from neg. feedback?

* Interesting since this form of regulation appears in ~~several~~ 3 systems
* Analysis is not ~~convincing~~ interesting — no laws, or conditions

In this report we introduce a mechanism of maintaining homeostasis (autoregulation) based on combination of a positive feedback and biphasic regulation. In this mechanism two molecular species interact in such a way that activation of the first of them affect activation of the second positively, while activation of the second species affect the activation of the first positively at low and negatively at high values. We demonstrate that a combination of biphasic response regulation with positive feedback leads to a possibility of homeostasis and limited graded response to variations of external parameters, possibly in a threshold manner. In addition, spike-like excitatory transient responses to perturbations of the system variables are possible. We then provide three examples of this sort of regulation in biological systems as diverse as transcriptional machinery, calcium channels and MAPK cascades. Correlation between the properties of biphasic response/positive feedback regulation and functional characteristics of these three biological systems is then discussed.

INTRODUCTION

Various biological systems from bacteria to humans are in constant need to both maintain homeostasis (processes leading to constant interior milieu) and be sufficiently sensitive to various external signals that may result in transient changes in homeostasis. One of the challenges is to maintain the concentrations of molecules constituting diverse signaling pathways at values optimal for signal reception prior to signaling event. If the concentrations are allowed to deviate significantly from the optimal levels, inadequate response may ensue. For example, whereas normally signal transduction mediated by mitogen-activated protein kinase (MAPK) can lead to cell proliferation, increased amplitude and duration of this signaling may arrest cell growth and lead to terminal differentiation (Marshall,

1995).

There is a distinction between two types of homeostatic regulation processes that often remains obscure. In one process a particular variable, such as the concentration of a regulatory protein, is maintained at a certain level in spite of some transient changes due to small external perturbations of this variable. In control theory such systems are said to be in a steady state with respect to this variable. We will refer to this form of regulation as *autoregulation*. In the second type of homeostasis the property of autoregulation is extended to include compensatory complete adaptation of a variable to persistent changes in some external parameter affecting the variable. This means that although the variable may change transiently in response to some parameter changes, mechanisms exist that compensate for this parameter changes in such a way that the variable always returns to the same steady state in the long run. We will refer to this type of regulation as *homeostatic adaptation*.

Homeostatic adaptation in living organisms has recently come to attention of control engineers, who proposed that integral negative feedback is a common feature of such regulation by analogy with controllers commonly used in design of various devices. One striking example of such regulation is adaptation to changing chemoattractant concentrations in bacterial chemotaxis. Similar processes may occur at physiological level. For instance, a rise in blood pressure is sensed by baroreceptors in carotid artery and aortic arch, which send the message about it to the medulla, from where a signal to reduce the heart rate then emanates. It is important to mention that stability or autoregulation is always assumed for systems regulated by homeostatic adaptation. Autoregulation is thus a more general feature of homeostatic systems.

In this report we introduce a mechanism of autoregulation based on combination of a positive feedback and biphasic regulation. In this mechanism two molecular species interact in such a way that activation of the first of them affect activation of the second positively, while activation of the second species affect the activation of the first positively at low and negatively at high values. Activation here is understood as the concentration of the active form of the species of interest. This form of regulation may seem counterintuitive, as autoregulation generally presupposes existence of negative feedback mechanisms restoring a variable to a steady state. However, we argue that biphasic regulation can provide both the necessary local negative feedback and limit the absolute values of the steady state. We show that this regulatory mechanism may be used for such important regulatory functions as calcium homeostasis, signal transduction through MAPK cascades and transcription regulation. We also demonstrate how autoregulation achieved through this mechanism is affected by the level of incoming signal providing a means for graded signal response and oscillations.

A THEORY OF BIPHASIC REGULATION COUPLED WITH POSITIVE FEEDBACK

In this section we treat mathematically some of the aspects of combination of biphasic regulation with positive feedback. We assume that there are two variables describing the system: $x(t)$ and $y(t)$, the evolution of which is described by the following equations:

$$\begin{cases} \dot{y} = f(S, x, y) \\ \dot{x} = g(x, y) \end{cases} \tag{13.1}$$

Here S stands for an external parameter that can affect the properties of f (it will be discussed later). We further assume that $y(t)$ is affected by $x(t)$ in a biphasic manner, whereas $x(t)$ is affected by $y(t)$ only positively (this latter assumption will be relaxed for the particular case of IP_3-sensitive Ca^{++} channels, as described below). These assumptions can be interpreted as follows. If x is fixed at a certain value, x_0, the first equation of [1] is assumed to ultimately evolve to $f(S, x_0, y_0) = 0$. If we now calculate the values of y_0 corresponding to all such x_0, we obtain a biphasic curve $y_0 = f^1(S, x_0)$. By analogy we can obtain the curve $x^* = g^{-1}(y^*)$, for all fixed y^*. This curve will have only positive slope. The curves $f^1(S, x_0)$ and $g^{-1}(y^*)$ are called null clines of (13.1). Plotting them in the same system of coordinates can be instructive about the behavior of the system. In our analysis we will have the obvious requirement that both null clines be attractive. That is for any fixed x_0, for example, the system will converge to the corresponding y_0 from any initial y, if given enough time. This requirement translates into the conditions:

$$\begin{cases} \frac{\partial f(S,x,y)}{\partial y} < 0 \\ \frac{\partial g(x,y)}{\partial x} < 0 \end{cases} \tag{13.2}$$

It can be demonstrated that these conditions and the character of null clines themselves (biphasic vs. positive slope curves) imply that any steady states of the system (intersections of the null clines) are stable, as long as they are on the "down-swing" part of the biphasic null cline. In the "up-swing" portion the conditions (13.2) still hold. However, an extra assumption needs to be made for stability of the steady state:

$$\left| \frac{\partial f(S, x, y)}{\partial x} \frac{\partial g(x, y)}{\partial y} \right| < \left| \frac{\partial f(S, x, y)}{\partial y} \frac{\partial g(x, y)}{\partial x} \right| \tag{13.3}$$

This is illustrated by a cob-webbing diagram (Figure 13.1), in which the consecutive states of the system are obtained by sequential "reflections" between the null clines of the system. It should be noted that cobweb diagrams may be misleading and are used here only for illustrative purposes.

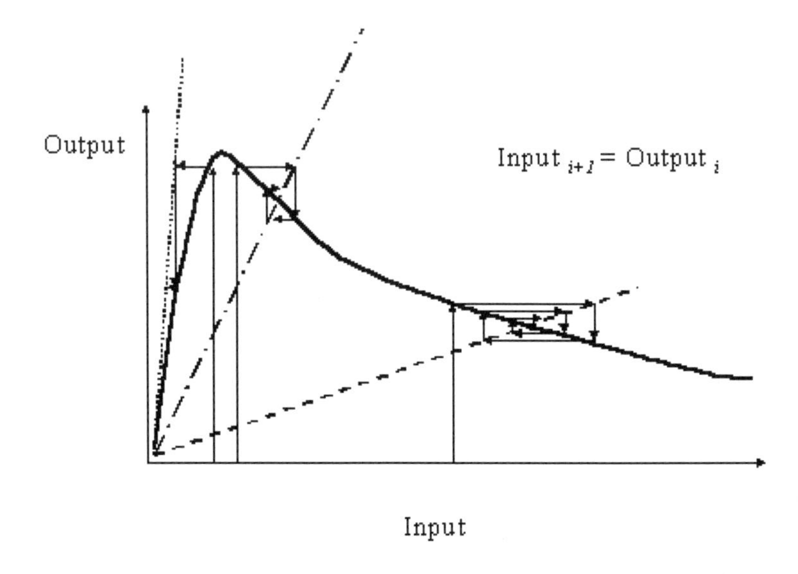

Figure 13.1 Three stable states defined by intersections of null clines of the system (13.1). The stability is illustrated by "cob-webbing', a method in which the trajectories are traced in discrete steps with each output taken to be the input for the next step. The result is a series of "reflections" from the system's null clines.

We are now ready to explore the properties of the system (13.1). First we will consider ways, in which the parameter S can affect the system. We will assume that the value of S does not influence the biphasic character of f, and only affects the relative amplitude of the null clines $f^1(S, x_0)$ and the position of the maximum. We also assume that dependence on S is monotonic everywhere. These assumptions are based in consideration of some particular examples of (13.1) below. We thus limit ourselves to analysis of two classes of behavior of (13.1) presented in Figure 13.2.

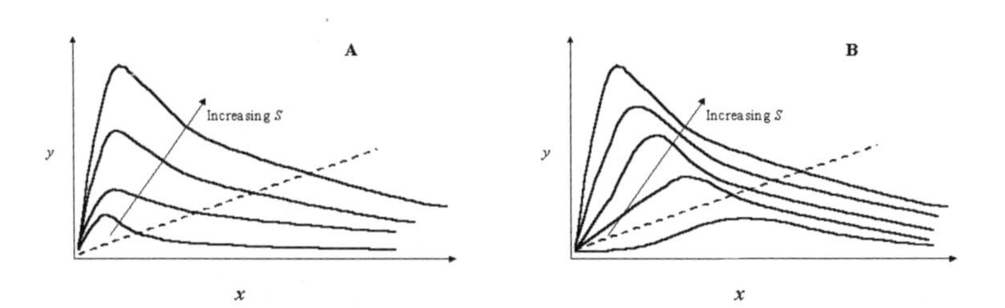

Figure 13.2 Increasing values of the external parameter S can lead to two types of changes in the null clines $y_0 = f^1(S, x_0)$. A. In the Type I behavior the null clines shift upward with maxima shifting toward higher values; B. In the type II behavior the null clines shift upward with maxima shifting toward lower values. It is evident that in the Type II behavior the steady state coordinates change from the low values around (0,0) to high values in a threshold way as a function of S.

In both classes an increase in the external parameter S leads to increasing values of f. In addition, however, in the first class increasing S also leads to positive shifting of the maxima of f, whereas in the second class it leads to negative shifting of the maxima of f.

It can be easily seen from Figure 13.2 that in class I behavior increasing values of S lead to gradual changes in the values of the steady state response of the system from very low to very high values. In the class II systems, however, the graded response sets in only at sufficiently high levels of S. There is thus a very sharp threshold for the steady state response as a function of S.

Since the "down-swing" part of biphasic response is indistinguishable form simple negative feedback, it is of interest to ask whether having biphasic response rather than just negative feedback can provide any advantages in regulation. It is evident from Figure 13.3, as well as from Figures 13.1 and 13.2, that biphasic response implies that the steady state values for y cannot exceed a certain limit. This is not true for simple negative feedback case, in which y can assume any value. Another potential consequence of having biphasic response regulation rather than negative feedback lies in the possibility of establishing sharp thresholds in response to variation of the external parameter S. Again, this dependence cannot be achieved if simple negative feedback rather than biphasic regulation is used.

The third potential advantage of having biphasic regulation is related to the kinetics of transient responses. Indeed, as illustrated in Figure 13.3, existence of the "up-swing" portion in the biphasic dependence curve leads possibility of a spike in the value of x in response to an external perturbation. As will be clear below, this possibility is important in regulation of Ca^{++} concentration. Having simple negative feedback precludes this possibility.

In conclusion, the analysis given in this section demonstrated that a combination of biphasic response regulation with positive feedback leads to a possibility of a limited graded response to variations of external parameters, possibly in a threshold manner. In addition, spike-like transient responses to perturbations of the system variables become possible. In the next section we provide three examples of this sort of regulation in diverse biological systems indicating how the properties outlines here serve to control autoregulation and response to external signals.

BIPHASIC REGULATION COUPLED WITH POSITIVE FEEDBACK IN BIOLOGY

Here we present three cases, in which biphasic regulation coupled with positive feedback is used for autoregulation and graded response to external signals on sub-cellular level. These examples are not exhaustive, nor are they meant to be. Their only function is to illustrate how this reg-

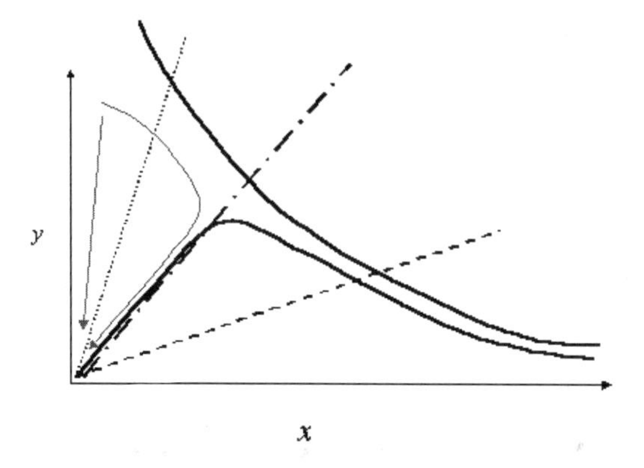

Figure 13.3 The main differences between the behaviors of the systems, in which positive feedback is combined with biphasic response (BR) vs. simple negative feedback (NF). The biphasic response and negative feedback null clines are shown as solid lines. Although for some positive feedback null clines (—) the system response is identical for BR and NF, for other possible null clines it is not. In the case of the positive feedback null cline intersecting BR null cline close to its maximum value (- · -) a spike-like response following perturbation (red trajectory) can be observed. No such excitory response is expected if the same steady state is given by combination of positive feedback and NF (green trajectory). Finally, combinations of positive feedback with NF, but not with BR, can result in unlimited values of y for the steady state (e.g., see high values for the null cline (. . .)).

ulatory scheme is utilized to achieve homeostasis and efficiently respond to stimulation.

TATA box binding protein and regulation of transcription

TATA boxes are elements commonly found in promoters of various highly expressed eukaryotic genes approximately 25-35 base pairs upstream of the start site. Mutations in TATA boxes frequently cause drastic reduction of transcription rate. It is now accepted that TATA boxes serve as a binding sites for the TATA binding protein (TBP), which, if bound, can provide (with some additional factors) nucleation sites for the assembly of general transcription machinery.

The current experimental evidence suggests that TBP interacts with TATA box and other regulatory factors as a monomer. At the same time TBP is capable of self-association and oligomerization. Oligomers formed from full-length protein include tetramers and octamers. Although it is not clear whether TBP oligomerization prevents association with DNA, the protein surface responsible for DNA binding apparently remains exposed in both oligomer forms. Therefore a possibility remains that TBP oligomers competitively interact with TATA box to prevent regulation by monomers.

The central role of TBP in transcription initiation suggests that its expression is tightly regulated. Investigation of the regulatory elements in TBP promoter reveals presence of a TATA box required for basal transcription and two control elements that bind another transcription factor: TBP promoter-binding factor (TBFP). One of the control elements, a higher affinity binding site located upstream of the TATA box, is activating and the other one, a lower affinity binding site located between the TATA box and the start element, is inhibitory for transcription activation(Huang and Bateman, 1997). The action of TBFP on the second site is negative presumably because it interferes with binding of TBP itself to the TATA box. Exposure of the TBP promoter to various amounts of TBFP resulted in a biphasic curve with the maximum at approximately 50 nM. A biphasic dependence is also predicted theoretically (Figure 13.4A).

As TBFP transcription is likely to be positively regulated by TBP (as part of the general transcription activation machinery), a combination of biphasic dependence of TBP transcription on TBFP concentration and positive dependence of TBFP transcription on TBP concentration emerges. Since, TBP transcription is also regulated by external signals involving Ras (Johnson et al., 2000), the biphasic dependence of TBP on TBFP can be modulated by the external parameters (designated as *S* above). Further simulations (Figure 13.4B) demonstrate that the biphasic dependence is of Type I, as classified above. In summary, the TBP-TBFP system described here is an example of the system [1] described in the theory section above.

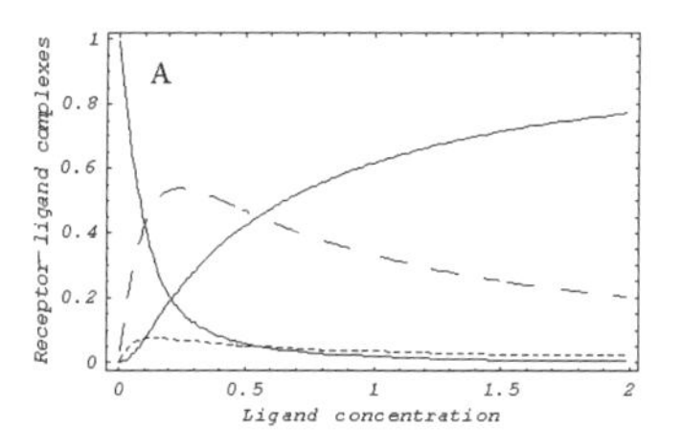

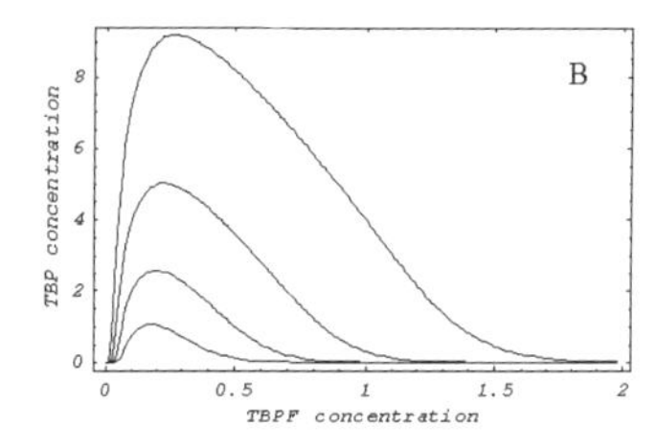

Figure 13.4 TBP expression is stimulated by TBFP in a biphasic manner. A. Existence of two binding sites on a receptor molecule (such as TBP promoter) for a ligand (such as TBP) leads to the illustrated dependence on the total ligand concentration. The complexes shown are empty and full receptor (solid lines), complex with the ligand present only at the high affinity site (- -) and only at the low affinity site. Since only the occupation of the high affinity but not the low affinity site signals activation, the biphasic dependence of response follows. The dissociation constants of 0.5 and 0.05 are assumed. The concentrations of complexes are normalized to unity. B. Dependence of TBP concentration on TBFP and Ras activation. Here competition of TBP and TBFP at the low affinity site was assumed (with equal dissociation constants of 0.05 vs o.5 for the high affinity site). Different rates of TBP production as a function of TBFP/TBP/DNA complexes give rise to the set of the curves. Type I dependence on the external signal is evident.

Since the expression of TBP needs to be maintained at limited levels even in the presence of activating signals, such those mediated by Ras, the use of biphasic regulation is best explained by property of the limited response identified above. We want to stress again that if simple negative feedback regulation was used instead, no upper limit on expression of TBP would be posed.

Regulation of the cytosolic Ca^{++} concentration

Ca^{++}-activated Ca^{++} release, shown to be important in a multitude of intracellular processes, is mediated by inosytol-triphosphate (IP_3) and Ryanodine Receptor (RyR) sensitive Ca^{++} channels in the endoplasmic reticulum. Of these the IP_3–sensitive channel, known as IP_3 receptor (IP_3R), has been studied to a larger extent (Keizer et al., 1995; Li, 1995). We thus restrict ourselves here to discussion of IP_3R. The opening probability of IP_3R is a biphasic function of cytosolic Ca^{++} concentration, arising from activation of the channel at low and inactivation at high Ca^{++} concentrations. In this case biphasic regulation stems from the presence of a high affinity activating and low affinity inhibitory Ca^{++} binding sites on each of the four receptor monomers. This opening probability dependence can be dramatically altered, both in terms of the position and absolute value of its optimum by varying IP_3 concentration (hence the name of the channel) (Mak et al., 1998) and ATP (Mak et al., 1999). Thus, increasing IP_3 leads to a positive shift in the position of the optimum and an increase in the maximum opening probability. This effect of IP_3 has been attributed to its ability to decrease allosterically the affinity of Ca^{++} to the inhibitory site, while the affinity to the activating site remains constant. Increase in ATP concentration shifts the position of the optimum negatively by decreasing the Ca^{++} affinity to the activating site. An important property of IP_3R is the inherent feedback of the output (probability of IP_3R opening) to the input (Ca^{++} concentration). The consequences of this feedback regulation are discussed below.

A range of intermediate of IP_3 concentrations has been reported to lead to Ca^{++} oscillations, while at the low and high IP_3 concentrations the levels are stable and depend on IP_3 in a graded way. As depicted in Figure 13.5, the reason for this behavior is the non-monotonic character of the feedback curve describing how the cytosolic Ca^{++} depends on the opening probability of IP_3R. It can be demonstrated that due to the non-linear dependence of the Ca^{++} pumps transporting Ca^{++} back to endoplasmic reticulum of the following (from (Li, 1995)):
$\frac{\alpha[Ca^{++}]^2_{cyt}}{\beta+[Ca^{++}]^2_{cyt}}$, the feedback dependence is a 3rd order polynomial of the shape presented in Figure 13.4. In the regions where the feedback becomes negative the steady states might lose their stability and oscillations may ensue. However, in the regions of true positive feedback all the steady

states are positive and the response is truly graded.

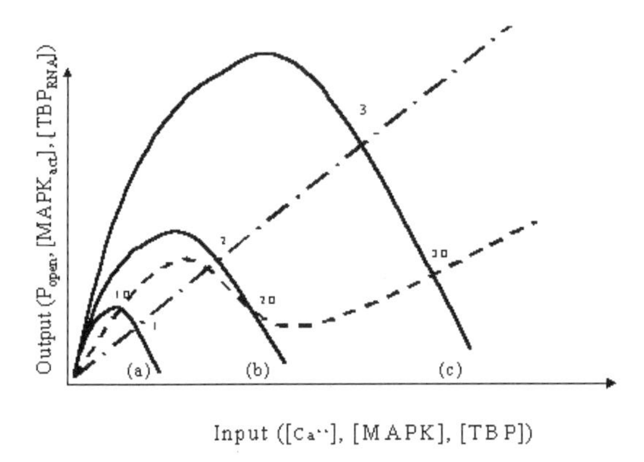

Figure 13.5 Optimal regulation in combination with a positive feedback can produce graded responses and help maintain homeostasis of signaling components. Solid curves correspond to optimal regulation of output by input. The level of a modifying component, such as IP_3 or MAPKKK activation increases from (a) to (c). The feedback curves correspond to monotonic positive feedback (- · -) for MAPK regulation and non-monotonic feedback (- -) for Ca^{++} regulation. The numbered intersection points correspond to steady states, 1,2,3, 1′, 3′ – all stable, and 2′ – unstable. The lack of stability at 2′ may cause oscillations around this steady state.

We have thus demonstrated again that cytosolic Ca^{++} regulation is an example of the system (13.1) above. The question of possible advantages of biphasic regulation vs. simple negative feedback can once again be posed. As in the case of TBP expression considered above, it is imperative for normal cell functioning to maintain low levels of Ca^{++} in absence of signaling. Indeed, Ca^{++} has been implicated in numerous regulatory functions including signal transduction, activation of muscle contraction, secretion and exocytosis and many others. Due to this important role in regulation of cellular function the cytosolic Ca^{++} levels (10^{-7} M) are maintained several orders of magnitude higher than the total cellular levels (10^{-3} M). The steep [Ca^{++}] gradient between cytosol and extracellular and various intracellular compartments is maintained by various means. A prolonged increase in intracellular [Ca^{++}] can be toxic. The benefit of an upper limit on the maximum output provided by a combination of biphasic response and positive feedback is thus obvious.

Another advantage provided by the system (13.1), as discussed above, is the possibility of the spike-like transient responses. This form of response is frequently observed following stimulation of excitable cells by various agents capable of transiently increasing cytosolic [Ca^{++}]. As can be seen from Figure 13.3, the spike-like response is possible only for certain values of the external parameter S. In case of Ca^{++} regulation, this parameter can be interpreted as IP_3. Our analysis thus predicts that the spike-like response, as well as oscillations, is possible only within certain ranges of cytosolic $[IP_3]$. This prediction fully confirmed by experimental data. A spike-like response would be impossible to implement if simple negative feedback was used instead of biphasic regulation.

Regulation of MAPK concentration

A MAP kinase (MAPK) cascade consists of three sequentially acting kinases (Garrington and Johnson, 1999). The last member of the cascade, MAPK, is activated by dual phosphorylation at tyrosine and threonine residues by the second member of the cascade: MAPKK. MAPKK is activated by phosphorylation at threonine and serine by the first member of the cascade: MAPKKK. The dual phosphorylation reactions occur in solution in a distributive manner, that is the two phosphorylation reactions are separated by full dissociation of kinase and its substrate. It has been shown theoretically (see Figure 13.6 for example) and experimentally that the distributive character of MAPKK and MAPK activation leads to a biphasic dependence of the signaling output on the concentrations of these kinases (Burack and Sturgill, 1997; Kieran et al., 1999; Sugiura et al., 1999). In simple terms this dependence results from saturation of the activating kinases (MAPKKK or MAPKK) by unphosphorylated substrates (MAPKK or MAPK, respectively) at high substrate concentrations, making second substrate phosphorylation unlikely.
In some systems MAPKK and MAPK expression can be up-regulated as a result of signaling in this pathway, thus creating a feedback on the level of the concentrations of the signaling components. In particular, increased expression of MAPK in pheromone and cell integrity pathways in response to activation of these signaling cascades has been documented by us (unpublished results) and others (Roberts et al., 2000). This may create a positive feedback from the activated MAPK onto the total level of MAPK in the system.

Such a system would again display all the characteristics of the system (13.1) above. As in the previous instances of biological regulation by a combination of biphasic response and positive feedback, limitation of the maximum response in the system seems to be a critical. Indeed, overexpression and overactivation of MAPK may lead to inappropriate responses, such as cell death or transformation. Simulations of MAPK activation reveal that the biphasic response may be of type II (Figure 13.7), which implies that there may be another benefit of using biphasic regulation – a switch like response. Indeed, switches in MAPK signaling have been described before with the mechanism different form proposed here (Ferrell and Machleder, 1998). We are now testing experimentally whether the threshold generating signal response mechanism proposed here is indeed present in yeast MAPK cascades.

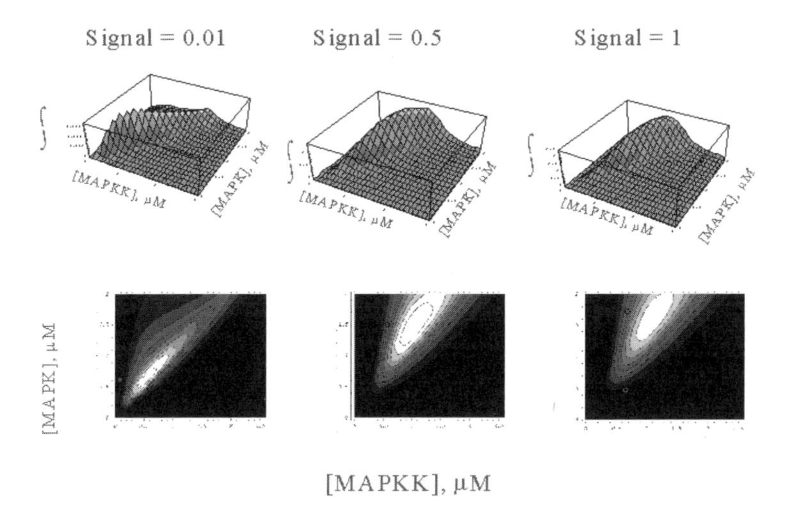

Figure 13.6 Biphasic dependence of MAPK activation on the total MAPK and MAPKK concentrations. Note that the time integral of MAPK dual phosphorylation over the first 2 min. rather than the steady state level of MAPK activation is shown. The red dots on the contour plots show locations of the maxima. [MAPK] is varied between 0 and 2 μM and [MPKK] is varied between 0 and 3 μM.

DISCUSSION

In this paper we describe a general mechanism of autoregulation and signal response arising from combination of positive feedback and biphasic regulation in interaction of two molecular species. We first showed that a mathematical analysis of a general description of such systems results in the property of autoregulation with limited graded response to variations of external parameters, possibly in a threshold manner. In addition, spike-like transient responses to perturbations of the system variables become possible. We then provided three examples of this sort of regulation in biological systems as diverse as transcriptional machinery, calcium channels and MAPK cascades.

Although a negative feedback regulation (which should be more properly viewed as a combination of mutual positive and negative feedback regulation between two molecular species) has long been demonstrated as means to maintain autoregulation in homeostasis, we show here that substitution of biphasic response for simple negative feedback can provide some regulatory advantage. In particular, existence of a maximum in biphasic response implies that there is an upper limit for a steady state activation of the system. Simple negative feedback, on the other hand, can result in a virtually infinite value for the steady state response of the system. In all three examples of biologic regulation discussed a limit in activation is of paramount importance. Overexpression of TBP or MAPK

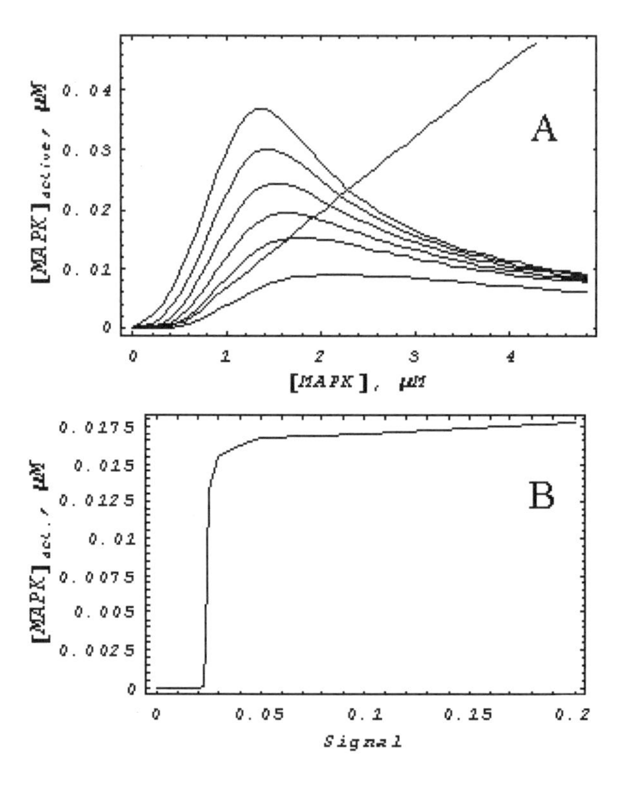

Figure 13.7 Simulations of the MAPK activation levels as a function of the total MAPK concentration. A. The biphasic curves combined with a positive feedback curve obtained from our simulations are shown for different activated MAPKKK concentrations (external signal). Type II character of the dependence on the external signal is evident. B. The steady state values obtained in A are plotted to reveal a sharp threshold dependence and switch-like signal reponse.

or high values of cytosolic [Ca^{++}] all can result in inappropriate cellular reactions, including apoptosis or transformation. Biphasic response regulation is thus critical for establishing and maintenance of appropriate activity values in homeostasis.

Another potential advantage of having biphasic response combined with positive feedback is possibility for spike-like responses or excitability. This property is heavily used in cell regulation by Ca^{++}, where Ca^{++}–induced Ca^{++} response is critical in such diverse processed as generation of the nerve impulse or muscle contraction. Here a balance between a very tight autoregulation in Ca^{++} homeostasis and fast and sensitive response is struck mainly due to biphasic regulation property.

Finally, we have demonstrated that combination of biphasic response may result in threshold response to variations in external parameter. Although this property remains to be demonstrated in the biological systems, we believe that it is likely to occur in regulation of MAPK cascade.

A set of experiments to verify this prediction in yeast MAPK cascades is currently under way at our laboratory.

Although the property of biphasic regulation has long been appreciated for it role in Ca^{++} oscillations, its functions in cellular regulation may go far beyond that. It is therefore of interest to further explore whether biphasic responses commonly observed in various biological systems are coupled to positive feedback as a general rule or only under certain circumstances. The analytical approach described here may serve to facilitate this task.

SIMULATIONS

Simulations for Figures 13.4-13.6 were performed within Mathematica® environment. The parameters for simulations in Figure 13.4 are given in the legend. The parameters used in obtaining Figures 13.5, 13.6 are the same as in (Levchenko et al., 2000).

ACKNOWLEDGEMENTS

This work was supported by Office of Naval Research Grant N00014-97-1-0293 and by a Jet Propulsion Laboratory grant to J.B. and P.W.S. (an investigator with the Howard Hughes Medical Institute), by a Sloan Research Fellowship to J.B., and by Burroughs-Wellcome Fund Computational Molecular Biology Postdoctoral Fellowship to A.L.

References

Burack, W.R. and Sturgill, T.W. (1997). The activating dual phosphorylation of MAPK by MEK is nonprocessive. *Biochemistry* 36:5929–5933.

Ferrell, J.E. Jr., Machleder, E.M. (1998) The biochemical basis of an all-or-none cell fate switch in Xenopus oocytes. *Science* 280(5365):895–898.

Garrington, T.P. and Johnson, G.L. (1999). Organization and regulation of mitogen-activated protein kinase signaling pathways. *Curr.Opin. Cell Biol.* 11:211–218.

Guillou, J.L., Nakata, H., and Cooper, D.M. (1999). Inhibition by calcium of mammalian adenylyl cyclases. *J.Biol.Chem.* 274:35539–35545.

Huang, W. and Bateman, E. (1997). Transcription of the Acanthamoeba TATA-binding protein gene. A single transcription factor acts both as an activator and a repressor. *J.Biol. Chem.* 272(6):3852–3859.

Johnson, S.A., Mandavia, N., Wang, H.D., Johnson, D.L. (2000). Transcriptional Regulation of the TATA-Binding Protein by Ras Cellular Signaling. *Mol. Cell Biol.* 20(14):5000–5009.

Keizer, J., Li, Y.X., Stojilkovic, S., and Rinzel, J. (1995). InsP3-induced Ca2+ excitability of the endoplasmic reticulum. *Mol. Biol. Cell* 6:945–951.

Kieran, M.W., Katz, S., Vail, B., Zon, L.I., and Mayer, B.J. (1999). Concentration-dependent positive and negative regulation of a MAP kinase by a MAP kinase kinase. *Oncogene* 18:6647–6657.

Levchenko, A., Bruck, J., and Sternberg, P.W. (2000). Scaffold proteins may biphasically affect the levels of mitogen-activated protein kinase signaling and reduce its threshold properties. *Proc. Natl. Acad. Sci. USA* 97(11):5818–23.

Li, Y.X., Keizer, J., Stojilkovic, S.S., and Rinzel, J. (1995). Ca2+ excitability of the ER membrane: an explanation for IP3-induced Ca2+ oscillations. *Am. J. Physiol.* 269:C1079–C1092.

Mak, D.O., McBride, S., and Foskett, J.K. (1998). Inositol 1,4,5-trisphosphate [correction of tris-phosphate] activation of inositol trisphosphate [correction of tris-phosphate] receptor Ca2+ channel by ligand tuning of Ca2+ inhibition [published erratum appears in Proc. Natl. Acad. Sci. USA 1999 Mar 16;96(6):3330]. Proc. Natl. Acad. Sci. USA 95:15821–15825.

Mak, D.O., McBride, S., and Foskett, J.K. (1999). ATP regulation of type 1 inositol 1,4,5-trisphosphate receptor channel gating by allosteric tuning of Ca(2+) activation. *J. Biol. Chem.* 274:22231–22237.

Marshall, C.J. (1995). Specificity of receptor tyrosine kinase signaling: transient versus sustained extracellular signal-regulated kinase activation. *Cell* 80(2):179–185.

Roberts, C.J., Nelson, B., Marton, M.J., Stoughton, R., Meyer, M.R., Bennett, H.A., He, Y.D., Dai, H., Walker, W.L., Hughes, T.R., Tyers, M., Boone, C., Friend, S.H. (2000). Signaling and circuitry of multiple MAPK pathways revealed by a matrix of global gene expression profiles. *Science* 287(5454):873–880.

Sugiura, R., Toda, T., Dhut, S., Shuntoh, H., and Kuno, T. (1999). The MAPK kinase Pek1 acts as a phosphorylation-dependent molecular switch. *Nature* 399:479–483.

14 Distinct Roles of Rho-kinase Pathway and Myosin Light Chain Kinase Pathway in Phosphorylation of Myosin Light Chain: Kinetic Simulation Study

Shinya Kuroda, Nicolas Schweighofer, Mutsuki Amano, Kozo Kaibuchi, and Mitsuo Kawato

Phosphorylation of myosin light chain (MLC) plays a key role in regulation of cell morphology, cell contraction and cell motility. The phosphorylation of MLC has been shown to be regulated by Rho-kinase pathway and MLC kinase (MLCK) pathway. However, due to the complex nature of signal transduction pathways and limitation of experimental approach, it remains obscure whether signal transduction pathways known at present can actually reproduce the whole process of phosphorylation of MLC. To address this issue, we built a diagram of phosphorylation of MLC and developed a computational kinetic model of phosphorylation of MLC based on kinetic parameters. Phosphorylation of MLC induced by thrombin has been experimentally shown to consist of two phases: the initial and the prolonged phases. The simulation reproduced the initial phase of the phosphorylation of MLC. The simulation also reproduced the activation of Rho-kinase pathway and MLCK pathway. However, the simulation failed to reproduce the prolonged phase, suggesting an existence of missing pathway responsible for the prolonged phase. The simulation revealed that both activation of MLCK and inhibition of myosin phosphatase by Rho-kinase are required for the sufficient phosphorylation of MLC. Thus, the kinetics simulation is a complementary tool to access the roles of signaling molecules and to predict a missing pathway(s) necessary to reproduce the whole process of phosphorylation of MLC.

INTRODUCTION

Cytoskeleton plays a crucial role in regulation of various cellular processes such as cell movement, cell morphology and cell contraction (Fishkind and Wang, 1995; Mitchison and Cramer, 1996; Stossel, 1993; Zigmond, 1996). The contractile force necessary for these processes is provided by actin-myosin interaction (Fishkind and Wang, 1995; Mitchison and Cramer, 1996; Stossel, 1993; Zigmond, 1996). In non-muscle cells and smooth muscle cells, phosphorylation of myosin light chain (MLC) has been shown to be a key reaction for regulation of actin-myosin interaction (Allen and Walsh, 1994). The MLC phosphorylation has been shown to be regulated by an increase of intracellular Ca^{2+} followed by activation of Ca^{2+}/calmodlulin-dependent MLC kinase (MLCK) (Allen and Walsh, 1994; Gallagher et al., 1997; Goeckeler and Wysolmerski, 1995; Somlyo and Somlyo, 1994). However, because the level of MLC phosphorylation is not always correlates to intracellular Ca^{2+} concentration, an additional mechanism which regulates the MLC phosphorylation has been proposed (Bradley and Morgan, 1987). Recent progress revealed that Rho-kinase, one of the effecters for the Rho small GTPase, regulates the MLC phosphorylation (Amano et al., 1996; Kaibuchi et al., 1999; Kimura et al., 1996). Rho-kinase phosphorylates MLC (Amano et al., 1996; Feng et al., 1999). Rho-kinase also phosphorylates myosin-binding subunit (MBS), a regulatory subunit of myosin phosphatase, and consequently inhibits the myosin phosphatase activity, resulting in elevation of phosphorylation level of MLC (Feng et al., 1999; Kimura et al., 1996). Thus, the MLC phosphorylation is dually regulated by MLCK and Rho-kinase pathways.

However, it still remains unclear whether agonist-dependent MLC phosphorylation can be reproduced by both the MLCK and Rho-kinase pathways. To address this issue, it is important to utilize the computational framework of kinetic simulation. We here built the computational simulation model of the MLC phosphorylation based on the kinetics parameters available in the literature or determined by the experiments by taking advantage of the recently developed program, GENESIS/kinetikit (Bhalla and Iyenger, 1999). The simulation reproduced the initial phase of the MLC phosphorylation, but failed to reproduce the late phase of the MLC phosphorylation, suggesting an existence of an unidentified pathway responsible for the late phase of the MLC phosphorylation. The kinetic simulation also suggests that the distinct role of MLCK and Rho-kinase in the regulation of the MLC phosphorylation. In addition, we attempted a re-interpretation of the effect of dominant active and negative form of Rho and Rho-kinase in the MLC phosphorylation.

METHODS

Block diagram of the MLC phosphorylation

To develop a kinetic simulation model of the MLC phosphorylation, we used thrombin (Essler et al., 1998; Goeckeler and Wysolmerski, 1995) as an agonist to induce the MLC phosphorylation. Addition of thrombin in some cell lines including endothelial cells has been shown to induce the MLC phosphorylation (Essler et al., 1998; Goeckeler and Wysolmerski, 1995). Signal of the stimulation of thrombin leads to activation of trimeric GTP-binding protein (G-protein), such as Gq and $G_{\alpha}12$. The activated Gq interacts with and activates phospholipase Cβ, resulting in production of inositol-1,4,5-phosphate (IP3) and diacylglycerol (DAG) (Bhalla and Iyenger, 1999). IP3 elevates intracellular Ca^{2+} and Ca^{2+} interacts with calmodulin (Bhalla and Iyenger, 1999). Ca^{2+}/calmodulin complex binds and activates MLCK and activated MLCK phosphorylates MLC (Allen and Walsh, 1994). On the other hand, the activated $G_{\alpha}12$ interacts with and activates guanine nucleotide exchange factor (GEF) for Rho, RhoGEF, resulting in the activation of Rho small GTPase (Hart et al., 1998; Kozasa et al., 1998; Majumdar et al., 1999, 1998). The activated Rho interacts with one of the effecters, Rho-kinase and activates it (Kaibuchi et al., 1999). The activated Rho-kinase phosphorylates MLC (Amano et al., 1996; Feng et al., 1999) and MBS (Feng et al., 1999; Kimura et al., 1996). Phosphorylation of MBS results in inactivation of myosin phosphatase and consequently elevates the MLC phosphorylation (Kimura et al., 1996). All kinetic parameters are previously reported (Bhalla and Iyenger, 1999) or shown in Table 14.1 and 14.2. The kinetic parameters in the Gα12 cascade is assumed to be the same as Gq (Bhalla and Iyenger, 1999).

Kinetic simulation

The MLC phosphorylation was simulated based on the following two biochemical reactions; the protein-protein (molecule-molecule) interactions and enzymatic reactions. The protein-protein interactions involve the interactions such as Ca^{2+}-calmodulin-MLCK and Rho-Rho-kinase. This reaction is given by the following formulation.

$$[A] + [B] \underset{K_b}{\overset{K_f}{\rightleftharpoons}} [AB] \tag{14.1}$$

In most cases, K_f and K_b were not available in the literature. Thus, based on the reported K_d values, the dissociation constant, K_f and K_b were calculated by following definition.

$$K_d = \frac{K_b}{K_f} \tag{14.2}$$

Enzymatic reactions involve the phosphorylation and dephosphorylation. This reaction is given by the following formulation of Michaelis-Menten.

$$[E] + [S] \underset{K_2}{\overset{K_1}{\rightleftharpoons}} [ES] \overset{K_3}{\longrightarrow} [E] + [P] \tag{14.3}$$

where E, S and P denote enzyme, substrate and product, respectively.

K_1 and K_2 values were not generally given in the literature. However,

K_3 can be calculated by experimentally shown K_{cat} value, given by the V_{max} divided by the concentration of the enzyme. K_m values are also generally reported. Thus, based on the K_m values and K_3 values, K_1 and K_2 values were calculated by following definition.

$$K_m = \frac{K_2 + K_3}{K_1} \tag{14.4}$$

Once the above parameters based on the kinetic values were set to reproduce the initial phase of phosphorylation of MLC, we simulated the following experiments. All numerical computations were performed with the kinetics library, which is an extension to GENESIS (Bhalla and Iyenger, 1999).

RESULTS AND DISCUSSION

To develop a computational kinetic simulation of the MLC phosphorylation, we first built a block diagram of the MLC phosphorylation based on the literature (Figure 14.1A). We used thrombin as an agonist to induce the MLC phosphorylation in this study. According to the literature, signal of thrombin can be divided into two pathways such as Ca^{2+}/calmodulin-activated MLCK pathways (Figure 14.1A, Figure 14.1B) and Rho/Rho-kinase pathway (Figure 14.1A, Figure 14.1C). The former pathway directly regulates the MLC phosphorylation (Allen and Walsh, 1994), while the latter pathway directly regulates the MLC phosphorylation and indirectly regulates it through the inhibition of myosin phosphatase (Kaibuchi et al., 1999). In the block diagram of Figure 14.1A, numbers of initial concentrations of molecules were 25, which were determined by the literature (Bhalla and Iyenger, 1999) or by our experiments (Table 14.1 and 14.2). Numbers of protein-protein interactions and enzymatic reactions were 33 and 19, respectively. All kinetic parameters such as K_d, K_m and K_cat on the above reactions have been reported (Bhalla and Iyenger, 1999) (Table 14.1 and 14.2) or determined by experiments (Table 14.2). Taking advantage of the kinetic parameters experimentally obtained, we built the kinetic simulation model of the MLC phosphorylation.

Using the kinetic parameters based on the literature and experiments, we tried to make the kinetic simulations to reproduce the experimental results of thrombin-induced MLC phosphorylation (Essler et al., 1998). It has been experimentally shown that the MLC phosphorylation induced by the stimulation of thrombin consists of at least two phases; the initial phase ($<$ 200 sec) and the prolong phase ($>$ 200 sec) (Figure 14.2A). In the kinetic simulation, the initial phase was induced in a thrombin dose-dependent manner, whereas the prolonged phase was not observed (Figure 14.2A). This result indicates that the initial phase, but not prolonged phase of the MLC phosphorylation could be reproduced. This result also

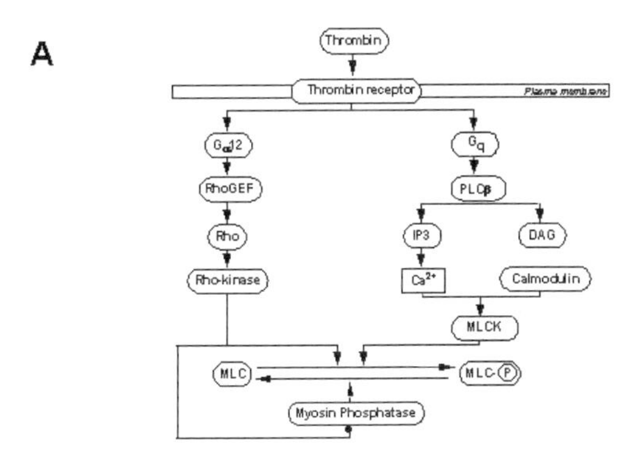

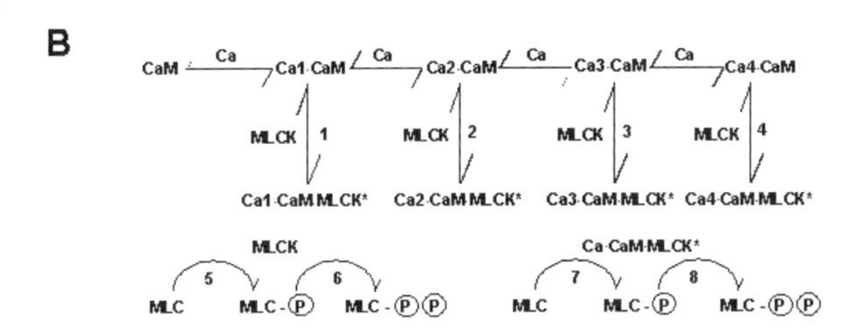

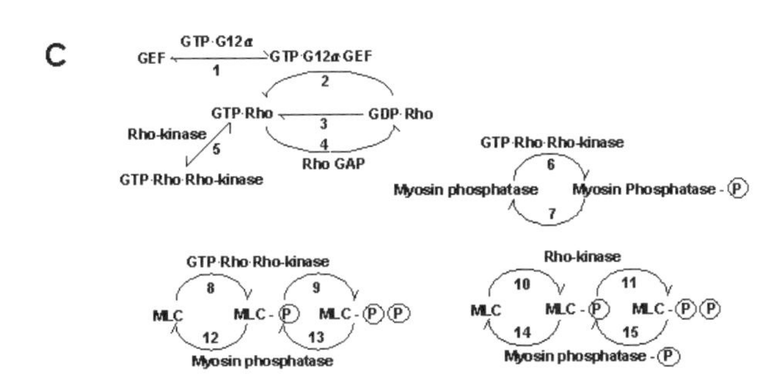

Figure 14.1 Block diagram of the MLC phosphorylation (a) Block diagram of the MLC phosphorylation. The stimulation of thrombin results in the elevation of the MLC phosphorylation through the two linear cascades, the MLCK pathway and the Rho-kinase pathway. Arrows and circles denote the stimulatory and inhibitory pathways, respectively. (b) Detailed block diagram of the MLCK pathway. Reversible reactions are represented as bidirectional arrows and irreversible reaction as unidirectional arrows. Enzymes are located on the middle of the segment. The numbers indicate the reactions whose kinetic parameters are shown in Table 14.1. Kinetic parameters of the interaction of Ca^{2+} with calmodulin are described elsewhere (Bhalla and Iyenger, 1999). The MLCK complex with either form of Ca^{2+}/calmodulin complex is assumed to show the same catalytic property (*). Abbreviations; CaM, calmodulin; Ca, Ca^{2+}. (c) Detailed block diagram of the Rho-kinase pathway. The numbers indicate the reactions whose kinetic parameters are shown in Table 14.2.

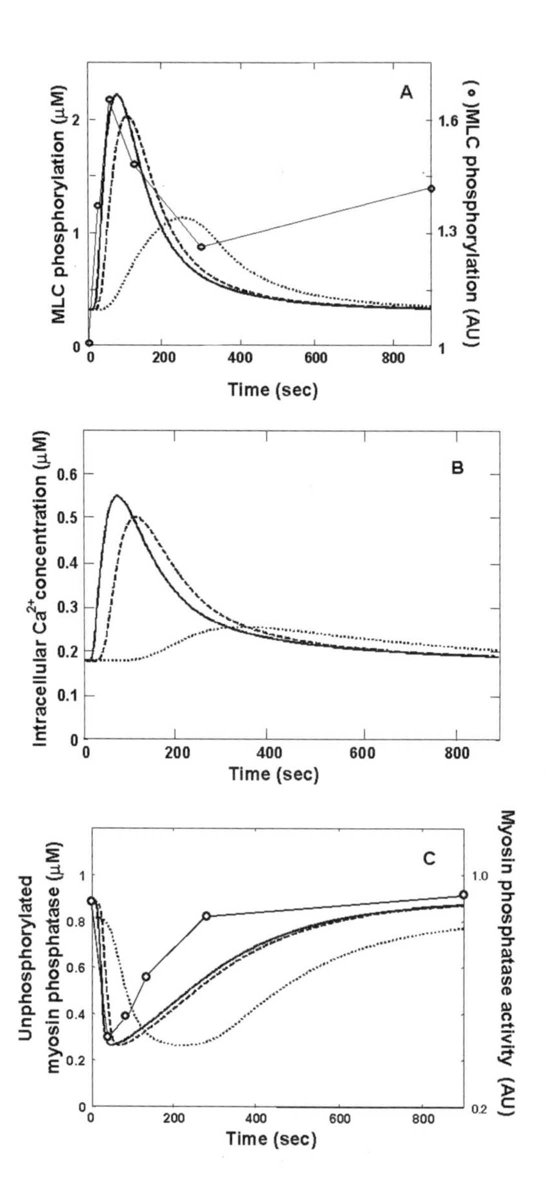

Figure 14.2 Thrombin dependent-MLC phosphorylation, the Ca^{2+} elevation, and the MBS phosphorylation. The simulation was run with various concentration of thrombin and the concentration of each indicated molecule was plotted. The concentration of thrombin; (--), 1 unit/ml; (–), 0.1 unit/ml; (···), 0.01 unit/ml. AU, arbitrary unit. (a) Time course of the MLC phosphorylation induced by thrombin. Circles are the MLC phosphorylation obtained by the experiments when 0.1 unit/ml of thrombin was used (Essler et al., 1998). (b) Time course of intracellular Ca^{2+} elevation induced by thrombin. (c) Time course of the phosphorylation of myosin phosphatase and of the myosin phosphatase activity induced by the stimulation of thrombin. Circles are the myosin phosphatase activity toward phosphorylase *b* obtained by the experiments when 0.1 unit/ml of thrombin was used (Essler et al., 1998).

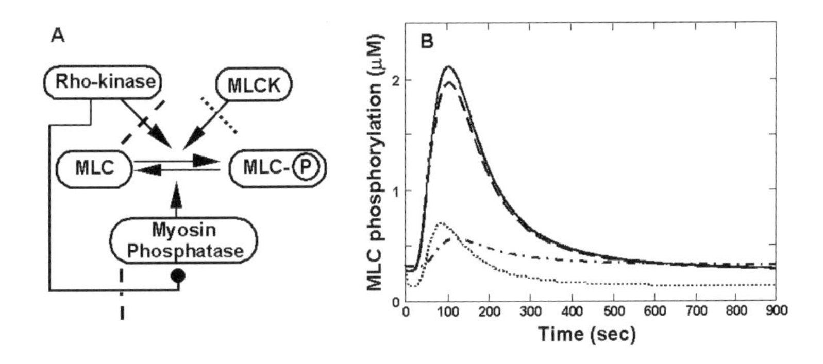

Figure 14.3 The role of MLCK, Rho-kinase, and myosin phosphatase in the regulation of the MLC phosphorylation. To explore the role of the indicated connections in the regulation of the MLC phosphorylation, the simulation was run under the condition where the indicated connections was deleted. The stimulation of thrombin 0.1 unit/ml was used. Deletion of the connections; (-), none; (..), MLCK-MLC connection; (- -), Rho-kinase-MLC connection; (-.-), Rho-kinase-MBS (myosin phosphatase).

suggests the possibilities that the thrombin-induced response of the two linear cascades, such as Ca^{2+}/calmodulin-activated MLCK pathways or the Rho/Rho-kinase pathway, could not be reproduced, or that an unidentified pathway(s) responsible for the prolonged phase of the MLC phosphorylation is missing in the block diagram. To exclude the former possibilities, we next asked whether the response of the two linear cascades to thrombin could be reproduced. We measured intracellular Ca^{2+} concentration in response to the stimulation of thrombin as an output of the Ca^{2+}/calmodulin-activated MLCK pathway (Figure 14.2B). In the kinetic simulation, intracellular Ca^{2+} concentration in response to the stimulation of thrombin showed the initial peak and reached the basal level 10 min after the stimulation (Figure 14.2B). This result is consistent with the earlier observation that the concentration of intracellular Ca^{2+} in response to the stimulation of thrombin shows transient initial peak and reached the basal level 10 min after the stimulation (Bahou et al., 1993; Goligorsky et al., 1989; Lum et al., 1992; Molino et al., 1997; Wickham et al., 1988). Next, we measured the unphosphorylated MBS concentration in response to thrombin as an output of the Rho/Rho-kinase pathway (Figure 14.2C), because the phosphorylation of MBS results in the inactivation of myosin phosphatase activity (Kaibuchi et al., 1999; Kimura et al., 1996). In the kinetic simulation, the unphosphorylated MBS concentration in response to the stimulation of thrombin showed the initial decreased peak and reached the basal level 10 min after the stimulation (Figure 14.2C). Although the decay in the kinetic simulation is faster than that obtained by the experiments (Figure 14.2C) (Essler et al., 1998), this result is consistent with the

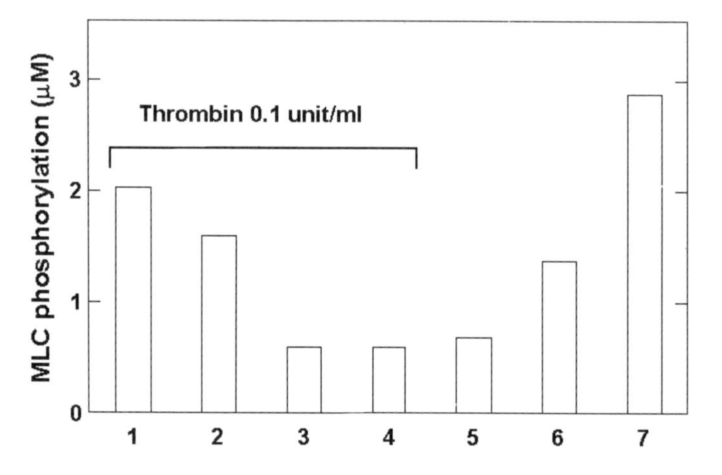

Figure 14.4 The effect of DA Rho and DN Rho on the MLC phosphorylation. The simulation was run with 0.1 unit/ml of thrombin stimulation. To examine the effect of DN Rho in the regulation of the MLC phosphorylation, DN Rho was introduced and the simulation was run with 0.1 unit/ml of thrombin (Lane 1 – 4). To examine the effect of DA Rho in the regulation of the MLC phosphorylation, the simulation was run under the conditions where the indicated concentration of DA Rho was added by holding its concentration as below without stimulation of thrombin (Lane 5 – 7). The maximum value of the MLC phosphorylation in the initial phase was plotted. Lane 1, none; Lane 2, 0.1 μM DN Rho, Lane 3, 1 μM DN Rho; Lane 4, 10 μM DN Rho, Lane 5, 0.036 μM DA Rho; Lane 6, 0.1 μM DA Rho; Lane 7, 1 μM DA Rho.

observation that the myosin phosphatase activity toward phosphorylase *b* transiently decreased and reached the basal level 10 min after the stimulation (Figure 14.2C) (Essler et al., 1998). These results indicate that the kinetic simulation reproduced the response of the two linear cascades to the stimulation of thrombin. Thus, it is likely that an unidentified pathway(s) responsible for the prolonged phase is missing in the current signal transduction pathways of the MLC phosphorylation.

The MLC phosphorylation has been shown to be dually regulated by the MLCK and Rho-kinase pathways (Figure 14.3A). It is intuitively difficult to understand how much each pathway, such as the MLCK- or Rho-kinase- dependent MLC phosphorylation, and the Rho-kinase-MBS-dependent dephosphorylation of MLC, contributes the whole process of the MLC phosphorylation (Figure 14.3A). To access the role of each pathway, we took advantage of the kinetic simulation by deleting each connection (Figure 14.3A, B). When the MLCK-MLC connection was deleted, the MLC phosphorylation dramatically reduced in response to the stimulation of thrombin (Figure 14.3A, B). By contrast, when the Rho-kinase-MLC connection was deleted, the MLC phosphorylation slightly

Table 14.1 Kinetic parameters in Ca^{2+}-calmodulin-MLCK pathway

Protein-protein interaction	K_d (μM)	Ref.
1	5	(Burger et al., 1983)*
2	4.5	(Burger et al., 1983)*
3	1.0^{-3}	(Adelstein and Klee, 1981)* (Burger et al., 1983)*
4	50	(Adelstein and Klee, 1981)* (Burger et al., 1983)*

Enzymatic reaction	K_m(μM)	K_{cat} (/sec)	Ref.
5,6	530	3.67	(Zimmer et al., 1984)
7,8	23.3	3.67	(Zimmer et al., 1984)

Molecule	Concentration (μM)	Ref.
MLCK	0.695	(Suzuki et al., 1999)
MLC	5	This study

* The K_d values were assumed from those toward δ subunit of phosphorylase kinase.

decreased (Figure 14.3A, B). When the Rho-kinase-MBS connection was deleted, the MLC phosphorylation dramatically reduced (Figure 14.3A, B). These results indicate that the direct MLC phosphorylation is mainly regulated by MLCK, but not by Rho-kinase, and that the inhibition of myosin phosphatase by the MBS phosphorylation by Rho-kinase is essential for the sufficient MLC phosphorylation in the kinetic simulation. Thus, the simulation is also useful to predict the role of molecule in a diverse pathway. However, it has recently been shown that MLCK and Rho-kinase play distinct roles in spatial regulation of the MLC phosphorylation (Totsukawa et al., 2000). Therefore, to conclude the role of each connection in the MLC phosphorylation, the distinct role of MLCK and Rho-kinase in the spatial regulation of the MLC phosphorylation should be further incorporated into the kinetic simulation.

MLCK activation is essential for the sufficient MLC phosphorylation in the kinetic simulation. On the other hand, the introduction of dominant active form of Rho (DA Rho), thought to be constitutively active in the cells, results in the sufficient MLC phosphorylation (Chihara et al., 1997; Gong et al., 1996; Kimura et al., 1996; Noda et al., 1995), although no evidence has been reported that the introduction of DA Rho leads to the elevation of Ca^{2+}, necessary for the activation of MLCK. If Ca^{2+} concentration is not affected by Rho, then how can Rho induces the MLC phosphorylation without MLCK activation? To address this issue, we examined the effect of DA Rho on the MLC phosphorylation without Ca^{2+} elevation by holding the concentration of DA Rho in the kinetic simulation (Figure 14.4A). The introduction of DA Rho induced the MLC phosphorylation in a dose-dependent manner (Figure 14.4A). When the concentration of DA Rho was the same as that induced by the stimulation of thrombin

Table 14.2 Kinetic parameters in Rho-Rho-kinase pathway

Protein-protein interaction	K_d (nM)	Ref.
1	5	(Kozasa et al., 1998)
5	50	This study

Enzymatic reaction	$K_m(\mu M)$	K_{cat} (/sec)	Ref.
2	0.02	0.1	(Hart et al., 1998)
4	2.83	0.9933	(Zhang and Zheng, 1998)
6	0.1	17.505	(Feng et al., 1999)
8,9	2.47	8.66	(Feng et al., 1999)
10,11	4.51	1.28	(Feng et al., 1999)
12,13	16.0	9.317	(Ichikawa et al., 1996)
14,15	58.1	1.95	(Ichikawa et al., 1996)

GTPase	k_1 /sec	Ref.
3	0.022	(Zhang and Zheng, 1998)

Dephosphorylation	k_1 /sec	Ref.
7	0.2	*

Molecule	Concentration(μM)	Ref.
RhoGEF	0.1	**
Rho	0.1	This study
RhoGAP	0.05	**
Rho-kinase	0.047	(Suzuki et al., 1999)
Myosin phosphatase	1.2	This study*

* The concentrations of the molecules were determined in MDCK cells.
** Assumption

(0.036 μM) as shown in Figure 14.2, MLC was not sufficiently phosphorylated. However, when the concentration of DA Rho was increased to 1 μM, MLC was phosphorylated to the similar extent to that induced by thrombin. When the concentration of DA Rho was held at 0.036 μM, about 10% of Rho-kinase was activated, whereas the concentration of the activated Rho was held at 1 μM, about 90% of Rho-kinase was activated (data not shown). Thus, it is likely that the overexpression of DA Rho induces the excess activation of Rho-kinase which is unlikely to occur under the physiological conditions.

We also examined the effect of the dominant negative form of Rho (DN Rho) in the kinetic simulation. DN Rho is thought to preferentially bind GDP rather than GTP and to inhibit the activation of endogenous Rho by the competition of the interaction with its activator RhoGEF. When DN Rho was introduced in the kinetic simulation, the MLC phosphorylation induced by the stimulation of thrombin was blocked in a dose-dependent manner (Figure 14.4B). Thus, the effect of DA Rho and DN Rho were reproduced in the kinetic simulation.

DA Rho and DN Rho have been shown to lead to the opposite effects on the MLC phosphorylation. This result seems to indicate that the MLC phosphorylation is regulated linearly by the Rho/Rho-kinase pathway.

However, in the kinetic simulation, both MLCK and Rho-kinase pathways are required for the MLC phosphorylation. In addition, as described above, under the physiological conditions, about 10% of Rho-kinase is activated by the stimulation of thrombin. Overexpression of DA Rho leads to the excess activation of Rho-kinase, resulting in the sufficient MLC phosphorylation, which is unlikely to occur under the physiological conditions. Therefore, it is generally important to measure how much population of the molecules are activated to determine whether the cascade is regulated by the linear pathway.

One of the potential advantages of the kinetic simulation is the application for drug design. Due to the complex nature of signal transduction pathways which regulate various cellular responses, it is intuitively difficult to know the consequence when a certain molecule or pathway is pharmacologically blocked or activated. However, using the kinetic simulation, we could predict the consequence when a certain molecule or pathway is pharmacologically blocked or activated. Therefore, this method will provide us with a novel approach for the efficient drug design.

Several molecules such as protein kinase C (PKC) are not included in the current model simply because the detailed molecular mechanism in the regulation of the MLC phosphorylation has yet to be shown. In addition, the distinct spatial roles of MLCK and Rho-kinase in the MLC phosphorylation has recently been shown (Totsukawa et al., 2000). Although the kinetic parameters to deal with the distinct localization are not available at present, this feature should be incorporated into the future model. It should be also emphasized that, even if the simulation can reproduce the experimental results, this does not deny the possibility that unknown pathways or unidentified molecules are required. In terms of the observation not reproducible by the simulation such as the prolonged phase of the MLC phosphorylation, a missing pathway(s) should be required to reproduce the observation. Therefore, the present kinetic simulation should not be regarded as a definitive model, rather as one of the complementary methods for exploring and predicting of the regulation of the MLC phosphorylation in addition to experimental methods.

References

Adelstein, R.S., and Klee, C.B. (1981). Purification and characterization of smooth muscle myosin light chain kinase. *J. Biol. Chem.* 256:7501–7509.

Allen, B.G., and Walsh, M.P. (1994). The biochemical basis of the regulation of smooth-muscle contraction. *Trends Biochem. Sci.* 19:362–368.

Amano, M., Ito, M., Kimura, K., Fukata, Y., Chihara, K., Nakano, T., Matsuura, Y., and Kaibuchi, K. (1996). Phosphorylation and activation of myosin by Rho-associated kinase (Rho- kinase). *J. Biol. Chem.* 271:20246–20249.

Bahou, W.F., Coller, B.S., Potter, C.L., Norton, K.J., Kutok, J.L., and Goligorsky, M.S. (1993). The thrombin receptor extracellular domain contains sites crucial for peptide ligand-induced activation. *J. Clin. Invest.* 91:1405–1413.

Bhalla, U.S., and Iyengar, R. (1999). Emergent properties of networks of biological signaling pathways. *Science* 283:381–387.

Bradley, A.B., and Morgan, K.G. (1987). Alterations in cytoplasmic calcium sensitivity during porcine coronary artery contractions as detected by aequorin. *J. Physiol.* 385:437–448.

Burger, D., Stein, E.A., and Cox, J.A. (1983). Free energy coupling in the interactions between Ca2+, calmodulin, and phosphorylase kinase. *J. Biol. Chem.* 258:14733–14739.

Chihara, K., Amano, M., Nakamura, N., Yano, T., Shibata, M., Tokui, T., Ichikawa, H., Ikebe, R., Ikebe, M., and Kaibuchi, K. (1997). Cytoskeletal rearrangements and transcriptional activation of c-fos serum response element by Rho-kinase. *J. Biol. Chem.* 272:25121–25127.

Essler, M., Amano, M., Kruse, H.J., Kaibuchi, K., Weber, P.C., and Aepfelbacher, M. (1998). Thrombin inactivates myosin light chain phosphatase via Rho and its target Rho kinase in human endothelial cells. *J. Biol. Chem.* 273:21867–21874.

Feng, J., Ito, M., Kureishi, Y., Ichikawa, K., Amano, M., Isaka, N., Okawa, K., Iwamatsu, A., Kaibuchi, K., Hartshorne, D.J., and Nakano, T. (1999).

Rho-associated kinase of chicken gizzard smooth muscle. *J. Biol. Chem.* 274:3744–3752.

Fishkind, D.J. and Wang, Y.L. (1995). New horizons for cytokinesis. *Curr. Opin. Cell. Biol.* 7:23–31.

Gallagher, P.J., Herring, B.P., and Stull, J.T. (1997). Myosin light chain kinases. *J. Muscle Res. Cell. Motil.* 18:1–16.

Goeckeler, Z.M., and Wysolmerski, R.B. (1995). Myosin light chain kinase-regulated endothelial cell contraction: the relationship between isometric tension, actin polymerization, and myosin phosphorylation. *J. Cell Biol.* 130:613–627.

Goligorsky, M.S., Menton, D.N., Laszlo, A., and Lum, H. (1989). Nature of thrombin-induced sustained increase in cytosolic calcium concentration in cultured endothelial cells. *J. Biol. Chem.* 264:16771–16775.

Gong, M.C., Iizuka, K., Nixon, G., Browne, J.P., Hall, A., Eccleston, J.F., Sugai, M., Kobayashi, S., Somlyo, A.V., and Somlyo, A.P. (1996). Role of guanine nucleotide-binding proteins–ras-family or trimeric proteins or both–in Ca2+ sensitization of smooth muscle. *Proc. Natl. Acad. Sci. USA* 93:1340–1345.

Hart, M.J., Jiang, X., Kozasa, T. Roscoe, W., Singer, W.D., Gilman, A.G., Sternweis, P.C., and Bollag, G. (1998). Direct stimulation of the guanine nucleotide exchange activity of p115 RhoGEF by Galpha13. *Science* 280:2112–2114.

Ichikawa, K., Ito, M., and Hartshorne, D.J. (1996). Phosphorylation of the large subunit of myosin phosphatase and inhibition of phosphatase activity. *J. Biol. Chem.* 271:4733–4740.

Kaibuchi, K., Kuroda, S., and Amano, M. (1999). Regulation of the cytoskeleton and cell adhesion by the Rho family GTPases in mammalian cells. *Annu. Rev. Biochem.* 68:459–486.

Kimura, K., Ito, M., Amano, M., Chihara, K., Fukata, Y., Nakafuku, M., Yamamori, B., Feng, J., Nakano, T., Okawa, K., Iwamatsu, A., and Kaibuchi, K. (1996). Regulation of myosin phosphatase by Rho and Rho-associated kinase (Rho-kinase). *Science* 273:245–248.

Kozasa, T., Jiang, X., Hart, M.J., Sternweis, P.M., Singer, W.D., Gilman, A.G., Bollag, G., and Sternweis, P.C. (1998). p115 RhoGEF, a GTPase activating protein for Galpha12 and Galpha13. *Science* 280:2109–2111.

Lum, H., Aschner, J.L., Phillips, P.G., Fletcher, P.W., and Malik, A.B. (1992). Time course of thrombin-induced increase in endothelial permeability: relationship to Ca2+i and inositol polyphosphates. *Am. J. Physiol.* 263:L219–225.

Majumdar, M., Seasholtz, T.M., Buckmaster, C., Toksoz, D., and Brown, J.H. (1999). A rho exchange factor mediates thrombin and Galpha(12)-induced cytoskeletal responses. *J. Biol. Chem.* 274:26815–26821.

Majumdar, M., Seasholtz, T.M., Goldstein, D., de Lanerolle, P., and Brown, J.H. (1998). Requirement for Rho-mediated myosin light chain phosphorylation in thrombin-stimulated cell rounding and its dissociation from mitogenesis. *J. Biol.Chem.* 273:10099–10106.

Mitchison, T.J. and Cramer, L.P. (1996). Actin-based cell motility and cell locomotion. *Cell* 84:371–379.

Molino, M., Woolkalis, M.J., Reavey-Cantwell, J., Pratico, D., Andrade-Gordon, P., Barnathan, E.S., and Brass, L.F. (1997). Endothelial cell thrombin receptors and PAR-2. Two protease-activated receptors located in a single cellular environment. *J. Biol. Chem.* 272:11133–11141.

Noda, M., Yasuda-Fukazawa, C., Moriishi, K., Kato, T., Okuda, T., Kurokawa, K., and Takuwa, Y. (1995). Involvement of rho in GTP gamma S-induced enhancement of phosphorylation of 20 kDa myosin light chain in vascular smooth muscle cells: inhibition of phosphatase activity. *FEBS Lett.* 367:246–250.

Somlyo, A.P., and Somlyo, A.V. (1994). Signal transduction and regulation in smooth muscle. *Nature* 372:231–236.

Stossel, T.P. (1993). On the crawling of animal cells. *Science* 260:1086–1094.

Suzuki, Y., Yamamoto, M., Wada, H., Ito, M., Nakano, T., Sasaki, Y., Narumiya, S., Shiku, H., and Nishikawa, M. (1999). Agonist-induced regulation of myosin phosphatase activity in human platelets through activation of Rho-kinase. *Blood* 93:3408–3417.

Totsukawa, G., Yamakita, Y., Yamashiro, S., Hartshorne, D.J., Sasaki, Y., and Matsumura, Y. (2000). Distinct Roles of ROCK (Rho-kinase) and MLCK in Spatial Regulation of MLC Phosphorylation for Assembly of Stress Fibers and Focal Adhesions in 3T3 Fibroblasts. *J. Cell Biol.* 150:797–806.

Wickham, N.W., Vercellotti, G.M., Moldow, C.F., Visser, M.R., and Jacob, H.S. (1988). Measurement of intracellular calcium concentration in intact monolayers of human endothelial cells. *J. Lab. Clin. Med.* 112:157–167.

Zhang, B., and Zheng, Y. (1998). Regulation of RhoA GTP hydrolysis by the GTPase-activating proteins p190, p50RhoGAP, Bcr, and 3BP-1. *Biochemistry* 37:5249–5257.

Zigmond, S.H. (1996). Signal transduction and actin filament organization. *Curr. Opin. Cell. Biol.* 8:66–73.

Zimmer, M., Gobel, C., and Hofmann, F. (1984). Calmodulin activates bovine-cardiac myosin light-chain kinase by increasing the affinity for myosin light-chain 2. *Eur. J. Biochem.* 139:295–301.

Index